Regulation, Market Prices, and Process Innovation: The Case of the Ammonia Industry

Westview Replica Editions

This book is a Westview Replica Edition. The concept of
Replica Editions is a response to the crisis in academic and
informational publishing. Library budgets for books have been
severely curtailed; economic pressures on the university presses
and the few private publishing companies primarily interested in
scholarly manuscripts have severely limited the capacity of the
industry to properly serve the academic and research communities.
Many manuscripts dealing with important subjects, often representing the highest level of scholarship, are today not economically viable publishing projects. Or, if they are accepted for
publication, they are often subject to lead times ranging from
one to three years. Scholars are understandably frustrated when
they realize that their first-class research cannot be published
within a reasonable time frame, if at all.

Westview Replica Editions are our practical solution to the
problem. The concept is simple. We accept a manuscript in camera-
ready form and move it immediately into the production process.
The responsibility for textual and copy editing lies with the
author or sponsoring organization. If necessary we will advise
the author on proper preparation of footnotes and bibliography.
We prefer that the manuscript be typed according to our specifications, though it may be acceptable as typed for a dissertation or prepared in some other clearly organized and readable
way. The end result is a book produced by lithography and bound
in hard covers. Initial edition sizes range from 400 to 600
copies, and a number of recent Replicas are already in second
printings. We include among Westview Replica Editions only works
of outstanding scholarly quality or of great informational value,
and we will continue to exercise our usual editorial standards
and quality control.

Regulation, Market Prices, and Process Innovation: The Case of the Ammonia Industry

Edward Greenberg, Christopher T. Hill, and David J. Newburger

Through the study of innovation in processes for the production of synthetic ammonia, the authors examine the effects of environmental and workplace regulations on business innovation in general. They present a history of ammonia production in the U.S., a survey of government regulation in the industry, and a model of process innovation that combines the economist's production function with the technical and practical concepts of the engineer.

Contrary to the widely held view that regulation has an unfortunate impact on business, the authors demonstrate that--at least in one industry--the economic factors of production have a measurable impact on innovation, while regulation does not.

Edward Greenberg is professor of economics at Washington University. Christopher T. Hill is senior research associate at the Center for Policy Alternatives, Massachusetts Institute of Technology. David J. Newburger is assistant professor of law at the Washington University School of Law.

Regulation, Market Prices, and Process Innovation: The Case of the Ammonia Industry

Edward Greenberg,
Christopher T. Hill,
and David J. Newburger

---

with the assistance of Thomas M. Helscher,
William V. Killoran, and Alan D. Norman

Westview Press **/** Boulder, Colorado

A Westview Replica Edition

This material was prepared with the support of the National Science Foundation Grant No. RDA75-23266. However, any opinions, findings, conclusions, and/or recommendations expressed herein are those of the authors and do not necessarily reflect the views of NSF.

Published in 1979 in the United States of America by
  Westview Press, Inc.
  5500 Central Avenue
  Boulder, Colorado 80301
  Frederick A. Praeger, Publisher

Library of Congress Catalog Card Number: 78-20661
ISBN: 0-89158-381-5

Printed and bound in the United States of America

To Joan, Sheila, and Toby

Contents

xi

Tables

xiv

Figures

Preface

Determining the extent to which environmental and workplace regulations have affected innovation motivated the research for this book. Our previous review of the literature* indicated that although many people held strong opinions on the subject, few had conducted empirical studies of the relationship between innovation and regulation. To approach the question, we believe that a suitable framework is necessary -- one that permits factors other than regulation to play their proper role, that portrays in a realistic fashion the legal and institutional environment within which decisions to innovate are made, and that reflects an engineering view of production processes and innovation. Accordingly, we worked as an interdisciplinary group, combining legal, engineering, and economic analysis.

The research is in the form of a case study of historical change in technologies used for the production of synthetic anhydrous ammonia. We placed particular emphasis on determining the ammonia industry's technological responses to input prices and to government regulation of the workplace and of environmental quality. We reviewed technical and legal literature on ammonia technology and regulation through December 1976. A tradition of publishing technical and economic details of processes allowed us to create time

---

*C. T. Hill, E. Greenberg, and D. J. Newburger, "A State of the Art Review of the Effects of Regulation on Technological Innovation in the Chemical and Allied Products Industries," Vol. II, Document #PB-243 728/AS, NTIS, Springfield, VA.

series of such inputs as fuel, feedstock, labor, and capital investment, as well as a history of U.S. ammonia production facilities, processes used, and capacities. A review of federal and state statutory and case law provided a history of the industry's regulation.

We developed a conceptual model of process innovation that combined the economist's production function with the engineer's concept of limited knowledge of technical options and constraints on plant capacity at any point in time. This led us to identify four kinds of ammonia process innovations: 1) those that reduce the required amounts of one or more inputs; 2) those that increase possibilities for substitution among inputs; 3) those that remove capacity constraints; and 4) those that introduce totally new approaches in an engineering sense.

Our major empirical findings were based on data for steam reforming of natural gas, a process widely used since World War II. We found that input factor coefficients respond to input prices with a 6-year lag, based on data for the period from 1947 to 1972. Regulation of workplace safety and health, although significant, could not be associated with process change. Regulation of environmental quality, primarily water pollution, has led only to minor process changes -- largely end-of-pipe pollution control.

Our study was conducted at the Center for Development Technology, Washington University (St. Louis). The project study team consisted of E. Greenberg, Professor of Economics; C. T. Hill, Associate Professor of Technology and Human Affairs and of Chemical Engineering; D. J. Newburger, Assistant Professor of Law; and T. M. Helscher, W. V. Killoran, and A. D. Norman, graduate students in Technology and Human Affairs, Law, and Economics, respectively.

This material was prepared with the support of National Science Foundation Grant Number RDA 75-23266. However, any opinions, findings, conclusions, and/or recommendations expressed herein are those of the authors and do not necessarily reflect the views of the National Science Foundation or the individuals and organizations whose aid is acknowledged below.

In addition to this volume, the National Science Foundation grant supported a master's thesis by Thomas M. Helscher that contains a great deal of detailed information on ammonia technology.

The thesis is titled "Process Innovation in the
Manufacture of Ammonia: A Case Study of Steam-
Reforming Natural Gas" (Department of Technology
and Human Affairs, Sever Institute of Technology,
Washington University).

Additional financial support, for which we are
grateful, was provided by Washington University's
Center for Development Technology, Department of
Economics, and School of Law.

We are pleased to acknowledge the helpful
research assistance of Michael Dogan, Christopher
Naish, Harvey Zar, and Ava Fried.

Typing for the project was speedily and
accurately performed by Donna Williams, Geraldine
Robinson, and Lynn Lawson. The final typescript
was prepared with exceptional skill and good humor
by Mrs. Williams with the assistance of Miss
Lawson. As usual, Emily Pearce was the perfect
Administrative Assistant.

We are grateful to the Monsanto Company of
St. Louis for the use of its research library and
to William A. Wilkenson for Assistance in its use.

We would like to thank the following persons
for helpful conversations and comments on our work:
A. E. Schaeffer of the Laclede Gas Company,
St. Louis, Missouri; H. Backes, W. H. Nelson,
E. Perry, D. Steinmeyer, and R. T. Webber of
Monsanto Company, St. Louis, Missouri; J. A.
Finneran and T. Czuppon of Pullman-Kellogg Inc.,
Houston, Texas; D. Collins, B. F. Day, G. Meyers,
and K. T. Johnson of the Fertilizer Institute,
Washington, D.C.; Daniel P. Boyd, Consultant;
Robert Nordhouse and Jeff Schwartz of the Sub-
committee on Energy and Power, House Committee on
Interstate and Foreign Commerce; W. Muir and
K. H. Jones of the Council on Environmental
Quality; J. A. Nelson, D. Davies, B. Diamond, and
E. Martin of the Environmental Protection Agency;
J. Proctor, J. Etherton, and G. Wrenn of the
Occupational Safety and Health Administration;
J. Douglas of the Tennessee Valley Authority;
E. B. Bagley of the United States Department of
Agriculture, Northern Regional Laboratories;
Maurice Pescitelli of the University of Illinois,
Urbana; and J. G. Myers and L. Nakamura of The
Conference Board, New York.

Finally, we would like to thank Professor
Robert P. Morgan, Director of the Center for
Development Technology, for his encouragement and
support of interdisciplinary research on technology
and human affairs.

1. Introduction

This study is an attempt to increase our understanding of technological change in the process industries. It is in the form of a case study of one such industry--anhydrous ammonia production--and therefore contains a substantial amount of information about processes in that industry. In many respects, however, ammonia production is typical of production of other chemical products, and of other industries which utilize similar technology. Thus it is believed that our approach and conclusions will be of more general interest.

Our particular concern is to examine the extent to which process innovation is affected by government regulations and such economic variables as input price changes. Thus, we take the viewpoint that innovations are, at least partly, amenable to explanation by appropriate methods of analysis and are not merely the products of random scientific discoveries. We emphasize regulation and economic variables for several reasons.

There exists substantial concern that government regulatory activity in the environmental and workplace areas has greatly increased production and capital costs. At the same time, others have argued that increased costs may be limited by the development and introduction of improved technology. Since federal regulation of these areas is relatively recent, we attempt to broaden our data base by a thorough review of regulation at all levels of government, going back as far as possible.

All numbered notes are grouped by chapter and appear in the "References" section at the end of the book.

Having set forth the regulatory environment, we attempt to understand its implications for innovation through interviews with people in industry and government, examination of professional journals, and engineering and legal analysis of the main issues.

With respect to economic variables, there is a considerable literature concerned with innovation. A number of variables have been suggested as influencing innovation, and we examine several of these. As a framework for our empirical findings we present a production model which seems to capture many aspects of production in process industries; this model yields a classification for process change which is illustrated by developments in ammonia production. A regression analysis of input process coefficients derived from the engineering literature is used to test the major hypotheses. Since our qualitative analysis shows that environmental and workplace regulations have not had a substantial effect on processes used in production of ammonia, we do not include these regulations in the regression analysis.

Ammonia is both an example of a general class of products and a product which has unique features. Some aspects of this uniqueness make it a particularly attractive industry for study; other aspects limit to some extent our ability to generalize. For the purpose of concentrating on technological change, we would like to have a standardized product which has been produced over a long time period using a number of processes. As will be seen below, ammonia satisfies these criteria. Moreover, ammonia utilizes a small number of inputs for which price data are readily available, and reasonably detailed process information may be found in the open literature. Since its production and transportation are somewhat hazardous, regulations have been in effect for a considerable period of time. Finally, ammonia is of substantial interest in its own right since its major use is as a source of nitrogen in fertilizer. At a time when American food production is of worldwide significance, the role of ammonia is critical. Furthermore, at a time when great attention is being paid to energy conservation and especially natural gas supplies, it is important to note that ammonia is a highly energy-intensive industry, using natural gas as both raw material and fuel.

2

For our purposes, however, ammonia has the
disadvantage that it has not been a primary target
of the recent concern with environmental pollution
problems. However, ammonia is a toxic gas which
is produced under very high pressures, and its
production, distribution and use pose significant
hazards to persons in the workplace environment.
Its production also poses important water pollution
problems. Moreover, urea, ammonium nitrate, and
nitric acid are objects of EPA concern; these
products are usually manufactured in the same
facility as the ammonia, often using processes
which are closely coupled to the ammonia synthesis
technology. We attempt to extend our data base
in the regulatory area by considering regulations
which were applied by state and local governments
in the past and by examining the implications of
current and proposed regulations.

In short, the ammonia industry may well find
itself in the center of a conflict between three
social objectives: an ample and inexpensive food
supply, a reasonable allocation of the nation's
remaining natural gas, and clean, safe workplaces
and natural environments. One possible resolution
of the conflict is process innovation which permits
nitrogenous fertilizer to be manufactured more
cheaply, more safely and cleanly, and with less
energy. As we discuss later, such process innovation has taken place in the past.

1.1   THE AMMONIA INDUSTRY

In this section, we provide general background
on the ammonia industry, including its role in the
economy and in the chemical industry, its history,
an overview of its manufacturing processes, and its
position regarding employee safety and environmental
protection regulation.*

In 1975, synthetic anhydrous ammonia ranked
third in production among the products of the U.S.

---

*A considerable literature exists for the ammonia
industry. Of special interest are the books by
Haber [2], Slack and James [3] and [4], Sauchelli
[5] and [6], Curtis [7], Ernst [8], Haynes [9],
Vancini [10], Matasa and Tonca [11], and a study
for the EPA by Development Planning and Research
Associates [12] .

chemical industry on the basis of weight, behind
sulfuric acid and lime. Its growth in the recent
past [1] has been extremely rapid:

TABLE 1.1
Annual Ammonia Production

Year	Production (million tons)	Year	Production (million tons)
1950	1.56	1973	15.47
1955	3.25	1974	15.80
1960	4.82	1975	15.78
1965	8.86	1976	16.71
1970	13.82	1977	17.40

    Current U.S. production takes place in approx-
imately eighty-three plants, nearly all of which
produce ammonia as primary product. A few plants
produce ammonia as a by-product of coke oven or
oil refinery operation. The industry at this time
is not greatly concentrated. The largest producers
and their capacities are presented in Table 1.2.
The 1963 four-firm concentration ratio for SIC
28191 (synthetic ammonia, nitric acid, and ammonium
compounds*) was 38%.
    The ammonia industry includes such profit-
making firms as Allied, DuPont, Union Carbide, and
Monsanto as well as cooperatives such as Farmland
Industries. The latter appear to be able to compete for at least three reasons: 1) tax advantages
for cooperatives, 2) locational advantages near
markets, and 3) the fact that complete plants can

---

*Changed for the 1973 Census to SIC 2873, Nitro-
genous Fertilizers (ammonia, nitric acid, ammonium
nitrate, and urea).

4

be purchased from major construction companies on a contract "turn-key" basis.

TABLE 1.2
Five Largest Firms in Ammonia Production in 1974 [13]

Firm	Annual Capacity (thousand tons)	Number of Plants
Allied Chemical	962	4
C. F. Industries	863	3
DuPont	780	3
Collier Carbon and Chem.	770	2
Chevron	745	3

The history of ammonia production may be traced back to the discovery of the importance of nitrogen in agriculture and the fact that many crops do not have the ability to utilize atmospheric nitrogen. Haber's history, The Chemical Industry, 1900-1930 [14], reveals that the first man-made source of fixed nitrogen was ammonium sulphate obtained as a by-product of crude coal gas. Note the role of environmental regulation in the industry's development as reported by Haber:

> At first, many of [the gas works] were so small that it was not worth their while to make the fertilizer, and the ammoniacal liquors were usually run to waste. By the 1890's some very large gas works had emerged and the pollution of streams was no longer tolerated.

Other man-made sources of fixed nitrogen in the years before the Haber-Bosch process included ammonia as a by-product of coke ovens, cyanamide (both as a fertilizer itself or as a source of

5

ammonia), and the electric arc process. It was the latter which originally drew Haber into the study of nitrogen fixation and ultimately led to the discovery of the Haber-Bosch process on which the present production is based.

For those interested in the sources of innovation, the Haber book notes that the possibility of food shortage and the resulting need for fertilizer was recognized before the turn of the century, and the demand for fertilizers was rising faster than the production of by-product nitrogen. In addition, there were difficulties with the natural sources, guano and Chilean sodium nitrate. Supply of the former was declining, and the supply of the Chilean product was in doubt both because of attempts to restrict supplies for higher profits and because of the possibility of war in which ammonia is used in explosives. The resulting great demand for man-made sources of ammonia induced several companies to devote research efforts in that direction. The payoff for the German company, BASF, was the Haber-Bosch process, for which Haber and later Bosch won Nobel Prizes.

The Haber-Bosch process was instituted on a large scale by BASF in Germany in 1913. The U.S. entered the era of direct synthesis at Muscle Shoals in 1918, but only very briefly as a late effort by the government to produce ammonium nitrate for use in World War I. This effort led to the rather unique involvement of the federal government in support of, and some say in competition with, private industry.

The government was involved first through the Fixed Nitrogen Research Laboratory, later a part of the USDA and an impetus for establishment of the USDA regional laboratories. Perhaps more significant, the government's efforts to use and/or dispose of the Muscle Shoals and related plants led ultimately to the formation of the Tennessee Valley Authority. The TVA to this day plays a role in developing techniques for blending and applying fertilizers and acts as a national and international focus for fertilizer information.

The U.S. government played a major role in the industry once again during World War II when ten new plants were constructed, six of which were based on a technology qualitatively different from the old coke-water gas process: steam reforming of natural gas. Moreover, the sale of most of these plants to private firms after the war had

6

important implications for market structure since the plants were not sold to the major prewar producers. [15], [16]

Although processes are discussed in considerable detail below, at this point it should be noted that the Haber-Bosch contribution was to develop a method for combining hydrogen and nitrogen to obtain ammonia directly. Much of the technological change to be discussed below is concerned with methods of obtaining hydrogen. These include electrolysis, coke-water gas, partial oxidation, and steam reforming. The two latter processes may be used, in general, with a variety of raw materials. The present U.S. industry relies primarily on steam reforming of natural gas. Other countries, where natural gas is less available, use naphtha as the feedstock in the steam reforming process. When natural gas is available at reasonable prices, the steam reforming process is capable of producing extremely low cost ammonia in efficient, large scale plants. However, it is very expensive to change feedstocks in the steam reforming process in contrast to the partial oxidation process. An interesting issue is the extent to which regulated prices of natural gas provided an incentive to perfect the steam reforming process to the relative neglect of alternatives. It is not surprising that current difficulties with natural gas prices and availability have awakened interest in processes which may be used with alternative raw materials, as well as in obtaining ammonia (and hydrogen) as by-products from materials contained in industrial wastes. Coincidentally, regulations designed to clean waste water from refineries, for example, have resulted in the adoption of processes which recover ammonia. [17]

As remarked above, regulation of natural gas prices may have helped shape the technology of the industry.* In addition, regulation of transportation has played a role in shaping the economics of the industry, particularly in its locational

----

*In the present study, this effect is accounted for indirectly through the use of natural gas prices which reflect actual prices paid by typical users. See Chapters 6 and 7 and the Appendix.

aspects.* It has been most economical for firms to locate near a source of gas and to ship ammonia, an arrangement which was facilitated by the innovation of shipment at low temperature in liquid form. An important concern has been that ammonia is a hazardous substance during shipping.

1.2 RELATIONSHIPS OF AMMONIA TO OTHER INDUSTRIES

The ammonia industry is closely related to a number of other products either as an intermediate in their production or through shared technologies. In this section we highlight a few of these relationships for the modern industry.

Ammonia, NH_3, is produced by the reaction of very pure nitrogen, N_2, and hydrogen, H_2, gases as follows:

$$N_2 + 3H_2 \longrightarrow 3NH_3$$

Ammonia can be oxidized in air to produce nitrogen dioxide, NO_2, which is absorbed in water to product nitric acid, HNO_3. This is the principal means for producing HNO_3.

Ammonia, a base, can be used to neutralize nitric acid to produce ammonium nitrate:

$$NH_3 + HNO_3 \longrightarrow NH_4 NO_3$$

Ammonium nitrate, a solid, is widely used as a fertilizer and as an industrial explosive.

Ammonia can also be reacted with carbon dioxide, CO_2 (a by-product of hydrogen production) to produce, through a two-step reaction, another solid fertilizer, urea:

---

*In the present study, impacts of transportation costs on plant location, plant size, and technology were not directly investigated. See Chapters 6 and 7.

8

$$2NH_3 + CO_2 \longrightarrow NH_4COONH_2$$
$$\text{(ammonium carbamate)}$$

$$NH_4COONH_2 \longrightarrow NH_2CONH_2 + H_2O$$
$$\text{(ammonium carbamate)} \qquad \text{(urea)}$$

Many modern integrated nitrogen fertilizer plants produce all four of these products: ammonia, nitric acid, ammonium nitrate, and urea in a single integrated complex. In fact, no stand-alone urea plants have been built because their economic operation depends upon use of the by-product CO_2. However, the available CO_2 is insufficient to use all of the ammonia to produce urea, so some ammonia or ammonium nitrate is nearly always marketed as well.

The production of ammonia requires large volumes of hydrogen gas, and this has been prepared from hydrocarbons or coal by such processes as partial oxidation, steam reforming, or the coke-water gas process. An intermediate product is synthesis gas, a mixture of hydrogen, carbon monoxide, carbon dioxide, and occasionally water, unreacted hydrocarbon, or nitrogen. Hydrogen and/or synthesis gas are key raw materials for many products including methanol, gasified coal, liquefied coal, and synthetic gasoline. Hydrogen and carbon monoxide are also keys to the reduction of iron ore. Many of these processes share common process elements, catalysts, and operating conditions with ammonia synthesis.

The early Haber-Bosch process represented the first large scale, high pressure process industry. As such, it fostered many developments in high strength alloys, pressure vessel manufacturing, instrumentation, pumps, and the like. These were subsequently employed in many other industries such as methanol synthesis, deep well drilling, polyethylene synthesis, and petroleum cracking.

1.3 PLAN OF THIS STUDY

Chapter 2 sets forth our view of technological change in the process industries by presenting a production model and a classification scheme for innovation. It is primarily concerned with economic variables and is placed in the perspective of attempts by economists to explain innovations in production. It concludes with several hypotheses.

Chapter 3 contains a description and discussion of the legal and regulatory environment in which the industry has been operating and the regulations which have been proposed. It considers workplace health and safety regulations, as initially developed by state and local authorities and more recently by OSHA, and air and water pollution regulations at various levels of government.

Chapter 4 describes the various processes which have been used for ammonia production in an historical framework.

Chapter 5 describes and analyzes the responses of ammonia process technology to regulations, based on several kinds of evidence.

Chapter 6 examines the dominant ammonia production process, steam reforming of natural gas, in detail, utilizing the model developed in Chapter 2. It also contains empirical tests of hypotheses relating economic variables and innovation.

Chapter 7 contains conclusions and interpretations, as well as recommendations for future research and suggestions for public policy.

The appendix describes the data which are used for the empirical analysis of Chapter 6.

References corresponding to numbered footnotes are listed, by chapter, at the end of the book.

2. A Framework for Studying Technical Change in Chemical Process Industries

In this chapter we describe our view of process innovation in general, present a production model within which technological change is assumed to take place, and discuss the implications of that model for process innovation.

2.1 GENERAL REMARKS ON INNOVATION

By process innovation we shall mean the first commercial-level use of a process, and innovative activity will refer to the research, development, and testing stages which precede an innovation. Since we believe it is fruitless, for purposes of this study, to distinguish between basic research, invention, development, and other stages in the conception and gestation of an idea, innovative activity will cover a rather wide set of activities. The main difficulty with differentiating between such stages is that an innovation, as defined above, usually consists of putting together several ideas, each of which has its own long and complicated history. Two aspects of the process of technical change must be examined for our purposes: innovation vs. substitution, and the nature of the lag between a stimulus to innovate and the resulting innovation.

We take the view that the traditional textbook production function considerably overstates both the ease of substitution between inputs and the extent of knowledge regarding production possibilities at input combinations far removed from input combinations presently in use or used in the recent past. [1] In particular, changes in factor proportions which appear to occur as a result of a change in input prices should not be

11

defined as (merely) substitution; rather, some attempt should be made to ascertain whether the second technology was actually available before the price change, or whether some research inputs were necessary to bring it about. This point is taken up below in connection with our empirical work. For complex chemical processes, for example, one does not have access to a "menu" of "off-the-shelf" alternatives which can be chosen costlessly depending upon factor prices.

With respect to the lag between an original idea and its first significant use, Rosenberg [2] reminds us that the implementation of an idea is as much an economic decision as a technical one. Thus, in some circumstances, a new idea may take a very long time to be implemented because input prices are not favorable or, if it is an innovation which requires considerable expenditure for new plant and equipment, the times may not have been propitious for new investment. These factors will complicate the estimation of the lag between stimulus and innovation.

2.2 ENDOGENOUS TECHNICAL CHANGE

        Implicit in the belief that regulation may affect innovation is the view that innovation is at least partly endogenous to the economy; i.e., that to a large extent, the activities of innovators are affected by the same types of signals as those to which other economic agents respond. This is not to deny that some amount of innovative activity takes place for purely personal reasons, or by virtue of governmental decisions which may be unrelated to considerations of profitability, or as the accidental by-product of an attempt to accomplish a different objective, or for a variety of other reasons. Furthermore, it does not imply that the innovative activity will be carried on by the firm which will actually use the process innovation: the activity may be carried on by specialized suppliers (e.g., firms which produce the catalysts used in ammonia production), by capital goods producers, or by design-engineering and construction firms. This view of endogenous innovation does assert, however, that statistical techniques should be able to isolate some degree of regularity in the relationship between measures of innovative

activity, innovation, and economic factors, making
the necessary allowance for random factors.*

Taking the usual economic model of the firm as
a starting point, expected profitability is the
touchstone for explaining firm actions. Let us
examine more closely the avenues through which
expected profitability might affect innovative
activity.

1. Expected demand increases: an industry
which is expected to grow is likely to attract
resources for the purpose of improving production
processes since cost savings will be greater, the
greater the volume of output. This factor is
associated with the work of Schmookler [3] and
Nordhaus [4].

2. Cumulative output: an industry which has
produced large quantities of a product may experience reductions in its required inputs. One
explanation is "learning by doing" in plant
operations; a second possible cause is a type of
"learning by doing" by the firms which produce
capital goods for the industry in question. That
is, as capital goods producers supply a large
number of units to an industry, they learn how to
improve the performance of their products.
Learning by doing in operations has been widely
studied [5]; learning by doing as related to
capital goods is noted by Arrow [6] and by
Schmookler [7]. Abernathy and Utterback [8] have
suggested a model for process innovation which
emphasizes interaction with the life cycle of the
product.

3. Input prices: the literature on induced
technical change takes the view that the direction
of technical change - whether it is capital or
energy saving, for example - may be influenced by
input prices. First suggested by Kennedy [9] in
connection with models of the entire economy, the
idea has been applied at the micro level by

_____

*We were unable to locate data on innovative
activity as measured by R&D expenditures, patents,
and the like. (But see Section 5.5). Since R&D
related to ammonia process technology has been
carried out by producers, equipment suppliers,
design and engineering firms, and the federal
government, no single source adequately measures
innovative activity.

Kamien and Schwartz [10], Nelson and Winter [11] and Binswanger [12]. In addition to influencing' the direction of technical change, changes in input prices may increase the payoff to process-oriented research in general, resulting in factor-saving innovation.

4. Regulations: limits on effluents or employee health and safety considerations may induce a firm to seek new processes which can meet the regulations, particularly if meeting them with existing technology will result in markedly higher costs. The process change may be rather marginal, such as end-of-pipe effluent treatment, or it may require a complete change in the process including the type of fuel and feedstock used. Examples of the effects of regulations and regulation-like activities may be found in the work of Rosenberg [13] and in that of Hill, et. al. [14].

A number of other economic factors affect innovation, of which the most important is probably market structure. Literature in this area has been recently reviewed by Kamien and Schwartz [15] and will not be further considered here. From this discussion it should be clear that there exist a number of economic variables which would appear to affect 1) the overall amount of innovative activity, 2) its distribution across industries, and 3) whether it emphasizes the saving of one or another factor of production.

The preceding has emphasized what Rosenberg has termed the demand for innovation [16]. He points out that the supply of scientific knowledge interacts with that demand to determine the innovation actually achieved. Although such issues in the case of ammonia are explored in some detail below, these factors are extremely difficult to disentangle in practice. Thus, metallurgical and other types of knowledge were required to produce the tubes that were needed for the high pressure reforming stage in ammonia production. But, to some degree, development of such tubes was undertaken with an eye on ammonia production. In other developments, ammonia may have benefitted from attempts to meet other objectives.

Figure 2.1 presents our view of the main factors determining process change. It depicts the exogenous influence of demand conditions, factor price changes, and government regulation on expected future profitability, with the latter two also influencing the direction of innovation. Expected profitability helps determine the amount

14

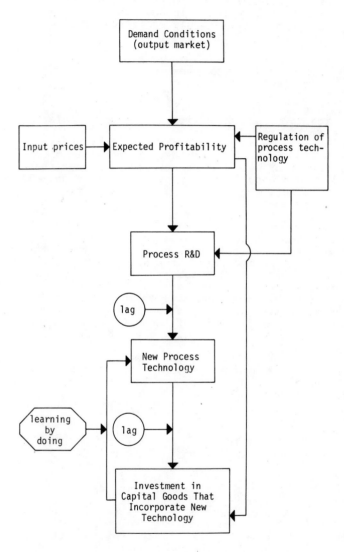

FIGURE 2.1. Factors Influencing
Process Innovation

15

of resources to allocate to process innovation and the amount of investment in new plant and equipment. In turn, the latter stimulates process R&D through the "learning by doing" route mentioned above.

2.3 A PRODUCTION MODEL FOR THE CHEMICAL PROCESS INDUSTRIES

Our efforts to model innovation in the chemical process industries, of which ammonia is an example, require that we first present a model which reflects an engineer's view of production. Production is assumed to take place within a finite set of well-defined processes rather than on a smooth neoclassical production function. In this respect, the model is a version of the "activity analysis" model. In contrast to the usual activity analysis model, however, some of the inputs are characterized by increasing returns to scale. Specifically, the general engineering view is that capital and labor (and perhaps other inputs) may be increased at a slower rate than output for a given process, while raw materials, energy, and other inputs are increased at the same or nearly the same rate as output. The economics literature on engineering production functions contains many of these points. In particular, see Moore [17], Chenery [18], and Teitel [19].

Assuming two inputs for simplicity, we designate the linear input by N and the nonlinear input by K. A first approximation to the production function for the ith process is:

$$Q_i = \min \left[\frac{N}{a_{ni,}} \quad \left(\frac{K}{a_{ki}} \right)^{1/b_i} \right]$$

where Q_i is output from the ith process, $K = a_{ki} Q_i^{b_i}$, and $0 < b_i \leq 1$. The coefficients a_{ni}, a_{ki}, and b_i completely characterize the ith technology in economic terms. For a given input of N and K, this function states that output will be the smaller of N/a_{ni} and $[K/a_{ki}]^{1/b_i}$; it does not permit tradeoffs between these two inputs within a particular process. However, unless $b_i = 1$, the ratio of inputs depends on the level

16

of output. In particular, K/N decreases with output along a given process for $b_i < 1$.

This model is displayed in Figure 2.2. The process lines represent combinations of N and K which satisfy $N/a_{ni} = (K/a_{ki})^{1/b_i}$. Since output is equal to the smaller of these two values, equality implies no redundant inputs. Output is proportional to the vertical axis, but the proportionality factor, $1/a_{ni}$, differs for each process. The exponent, b_i, is usually found to be approximately .6 or .7 if K represents plant and equipment. Because of the increasing returns to K when $b_i < 1$, it will generally not be profitable to produce output by using more than one process, so that many combinations of inputs will not be utilized when a small number of processes are available.

Subject to modification concerning substitution possibilities noted below, the model we have in mind is of the "putty-clay" variety. That is, although there is a choice of coefficients before a process is installed, the coefficients are fixed thereafter.* This model permits a much wider scope for choice of coefficients than the limited set actually available for producing most chemical products. Our view is that, at any given time, there are only a finite number of processes available, and in most cases that number is rather small.

Two modifications must be made to this model to make it conform more closely to engineering practice. First, there are upper and lower limits to the output possible from a given process at any point in time. Second, a certain amount of substitution between inputs, within a process, is permitted. This type of substitution permits a degree of variation in a_{ni} and a_{ki}, but these variations will be small relative to the differences in coefficients between processes.

---

*Plant and equipment is measured in dollars, although the dollars may purchase rather different specific items for different processes.

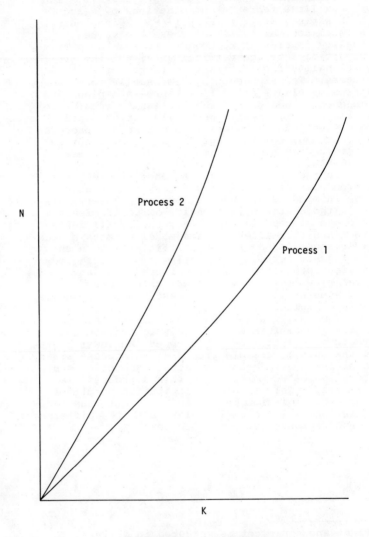

FIGURE 2.2. Process Model

18

Bounds on Process Size

   The existence of upper and lower bounds on
the scale of a particular process may be explained
by a combination of technical and economic reasons.
To some extent the bounds are a simplified method
of indicating a sharp nonlinearity in the price of
capital equipment when units of nonstandard size
must be ordered. This may make the cost of the
resulting input combination so high that it would
not be used in the short run. In other cases,
building equipment at very small (or very large)
scale is beyond the state of engineering capability.
For example, centrifugal compressors require a
minimum scale of output unless wasteful gas
recirculation is employed to build artifically high
throughput.*
   Upper bounds on process size are of particular
interest since, in the presence of increasing
returns to scale, one might expect that plant size
would increase to the point that just one plant
produces all the output. A number of factors
constrain this growth. First is the high cost of
using nonstandard sizes, or indeed, the cost of
producing made-to-order equipment much larger than
that which is ordinarily produced. Costs of production, shipping, and installation of such
equipment may rise faster than scale; we treat
this as a constraint rather than a nonlinearity
in the price of the input. In ammonia production,
a process which utilizes extremely high pressures,
the vessels must be very thick and strong. Moving
such vessels to a desired site may be beyond the
capacity of transportation systems, and on-site
fabrication may be very expensive.
   Second, there is the problem of reliability.
If a malfunction occurs in one of two 1500 tons
per day (tpd) plants, for example, only half of
the production is lost compared to the loss from
the shutdown of a single 3000 tpd plant. A
related factor is the loss in production during
start-up time, although the general impression
in the literature is that start-up experience with

----

*Early applications of centrifugal compressors in
ammonia plants were considered only for the final
pressure boost to the synthesis unit. The throughput at this point is several times the output
level due to the high degree of recycle of input
gas employed. See Section 4.4.

the large ammonia units has been good. [21] Our
impression is that producers seem unwilling to
invest in plants greatly larger than existing
plants. As experience accumulates on the largest
plants, however, firms become more willing to
increase plant size further.

A third consideration is the ability to
operate the process at less than full capacity,
along with the sensitivity of costs to less than
full capacity operation if such operation should
become desirable. Average costs at less than full
capacity are, of course, sensitive to the
relationship between the fixed and variable costs
of a process. For ammonia production, we find
that centrifugal compressors require a minimum
throughput to operate at all, so that operation at
less than about 65% capacity is impossible.

Market size and transportation costs may set
the most significant limits on maximum plant
size. [22] Whereas the unit production costs
implied by our model decrease, the rate of decrease
slows. Thus the gains from increasing scale, for
a given process, become less and less important.
At the same time, if the plant's production will
be a large portion of the total market or if
customers will be very large distances away from
the plant, the net price received by the producer
may fall. The marginal revenue may fall faster
than marginal cost, yielding an economic upper
limit to production. This boundary is sensitive
to the size of the market for ammonia, which is
determined primarily by agricultural commodity
prices, by costs of storage and transportation, by
prices of other farm inputs and by the weather
each year. These factors are important in plant
location and scale decisions. Typical alternatives
are to build a 2000-tpd plant at the Gulf Coast,
or a 1500-tpd Gulf Coast plant and a 500-tpd
plant in Kansas. It is necessary to trade off the
savings in unit production costs against the
increased transportation costs, taking into account
the differences between land and water transportation rates, and handling costs. In any event,
the largest operating plant in 1976 had a capacity
of 1500 tpd, which was only 2.6% of total U.S.
capacity.

Input Substitution Within A Process

The second modification to our increasing returns activity analysis model is that we permit some degree of substitution among inputs along a given process ray. For example, it may be possible to use less feedstock and more fuel to produce the same output from fixed capital equipment. The extent of this substitution is limited by physical and chemical laws, as well as by legal restrictions and by equipment limits in the short run. Thus, the production of one ton of ammonia by steam reforming of natural gas requires a minimum of just under 16 million BTUs of natural gas in order to obtain the hydrogen contained in the ammonia. It is possible to conserve on fuel natural gas by using another energy source, but the feedstock input is bounded from below so long as one uses steam reforming alone. Moreover, this calculation assumes a complete conversion; under actual operating conditions, something over 20 million BTUs are needed as feedstock. A general framework for the relation between inputs and outputs in chemical production has been developed by Rudd [23]: inputs are desired for their molecular composition, energy content, and cooling ability; inputs also generate by-products which may contaminate the process or create air and water pollution. Such effects are inherent in the chemistry of the process.

One kind of substitution between market inputs is what is called the "make or buy" decision. Depending upon local conditions, a firm may find it profitable either to buy untreated water for its boilers and do its own treatment or to buy treated water from a utility. Similarly, a firm may either buy all of its cooling water requirements from a utility or it may recirculate cooling water, buying only make-up water and supplying the necessary cooling towers and electric power to operate them. In both cases the firm is buying inputs for the purpose of cooling; however, in terms of market behavior, this may show up as large purchases of treated water on the one hand, or smaller purchases of water and greater expenditures on capital and electricity on the other. The decision will be influenced by such conditions as local prices and regulations.

A second type of substitution occurs in the form of substitutions of subprocesses within a particular process. For example, one step in the steam reforming process is the removal of carbon monoxide and carbon dioxide from the synthesis gas. A number of processes have been developed for this purpose, differing somewhat in energy and capital requirements. These are discussed in Chapters 4 and 6.

The final kind of substitution we consider is the possibility of changing operating conditions within a particular process, resulting in moderate changes in coefficients. The most important example in ammonia production is the choice of pressure at which the steam reformer is operated; coefficients of capital, fuel, and cooling water for producing one ton of ammonia are affected. This example is also treated in detail in Chapters 4 and 6.

Engineering production processes may be characterized in dimensions other than their primary inputs and outputs. They are associated with differences in at least the following: effluents, requirements for skilled labor, reliability, safety, adaptability to use of alternative feedstocks, and sensitivity of costs to less-than-full-capacity production. These characteristics are not easily represented in a simple two-dimensional production function diagram. As an illustration of our model, however, we plot the curves for two variants of the steam reforming process which are similar in most respects other than capital and electricity usage. Figure 2.3 contrasts steam reforming with reciprocating compressors (process P_1) to steam reforming with centrifugal compressors (process P_2), using industry data. The curves are drawn to indicate the limits to capacity experienced by the industry in the mid 1960s. Capital is measured in millions of dollars since the equipment used in the processes is qualitatively different. There is a small difference between P_1 and P_2 in natural gas consumption, although not enough to account for the difference in electricity consumption. Process P_1 generates low pressure steam, the value of which depends greatly on local conditions.

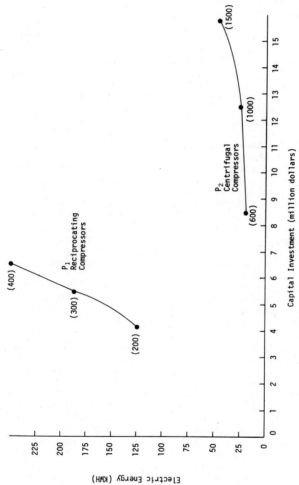

FIGURE 2.3. Electric Energy and Capital Inputs for Two Stream Reforming Processes (output in tpd in parentheses). Source: [24]

23

2.4 IMPLICATIONS OF THE PRODUCTION MODEL
FOR PROCESS INNOVATION

The production model discussed permits us to
classify innovations into four main categories.
At this point we shall identify these categories
and relate them to the literature on innovation.
The concepts are applied to innovation in ammonia
production in Chapter 6.

The most obvious type of innovation is one
which reduces the amount of one or more inputs
required for the ith process; that is, one which
reduces the a_{ni}, a_{ki}, or b_i. Such innovations may
result from improvements in catalysts or from
improvements in the physical arrangement or
efficiencies of equipment.

A second type of innovation is an increase in
the substitution possibilities around a particular
process, for example, substitution of energy for
capital. Whether any particular implementation of
such substitution is an innovation or merely a
substitution as the term is used in the usual
theory of production depends on the extent of
research and development necessary to implement it.
In the ammonia case, Kellogg engaged in a considerable research effort to determine the effects
of varying pressures on the output of reformers
in order to permit the substitution of capital for
energy [25], and their work is considered to constitute an innovation.

A third type of innovation is concerned with
attempts to weaken the capacity constraints at
both upper and lower limits. At the upper limit,
weakening a constraint will reduce unit costs, on
the assumption that the production relationship
holds at greater sizes with the same coefficients
which characterized smaller outputs. Production
beyond existing upper limits, as noted above, may
be associated with difficult engineering problems.
Expansion of output by moving along a process
curve may not be possible because of additional
stresses put on equipment. The experience in
ammonia production seems to be a gradual increase
over time in the size of plant used in any process.
This suggests a gradual learning about how a plant
behaves at a given size before attempting larger
sizes. (However, the shift to centrifugal compressors was associated with a jump in capacity.)
In addition, attempts to reduce the minimum scale
at which a process may be operated may also be
worthwhile because of local conditions which both

24

justify a smaller output than may be produced by a
given process and make a process attractive because
of relative input prices. Thus, it may be desirable to reduce the capacity needed to permit the
use of centrifugal compressors in the steam reforming process because they require far less
electric power than reciprocating compressors.
Local conditions may require outputs less than
600 tpd, but have relatively high electric prices
which would make centrifugals attractive.

A fourth type of innovation is one which
results in an entirely new process in the engineering sense. All the process coefficients may
be very different, entirely different inputs may
be used, or new physical or chemical principles
may be employed. In contrast to production of
ammonia from coal, steam reforming of natural gas
was such an innovation, as was the shift from
propellor to jet aircraft.*

The first two types of innovation noted above
are similar to those utilized in the innovation
models of Nelson and Winter and Binswanger. These
models are concerned with movements of input coefficients within a region close to their initial
values. However, since these models neither
incorporate constraints nor permit increasing
returns to scale there is no innovation along the
lines of the third type discussed. The dissertation of Levin recognizes this type of innovation.
The Nelson and Winter approach is based on alternative coefficients generated by a random
process -- perhaps because basic scientific discoveries are rather unpredictable -- and employs
a modified profitability test for their adaptation.
Binswanger studies the generation of alternative
processes in a deterministic model in which
research expenditures may be allocated to change
coefficients. The Nelson and Winter model assures
the generation of points near existing points by
appropriate specification of the random process.
A specification of the R&D cost of changing coefficients which assured relatively small period-to
period changes would make the approaches consistent
with the same data.

----

*Research efforts to produce nitrogen-fixed
(ammonia-producing) bacteria which are symbiotic
with corn through DNA manipulation, or gene
splicing, is innovative activity of the fourth
type. [26]

Kamien and Schwartz have put forth a variety
of models which permit the introduction of innovations that either cause increases in output for
the same inputs or expand differential input substitution possibilities. The returns from research
are deterministic in their models, and fairly
general R&D cost functions may be used. An
attempt to model search activity for improving
production processes has been made by Roth [27].
He distinguished between investigating known
processes and spending resources to learn the properties of combinations of subprocesses not previously used. The existence of subjective pro-
bability distributions of returns to the various
activities is assumed. It may be that the type
of engineering data used in the present study can
provide information about the frequency distribution of changes in coefficients to be used in a
Nelson-Winter type model. Although these data may
be helpful for empirical implementation of such
induced innovation models as those of Binswanger
and Kamien and Schwartz, the absence of research
and development expenditure data on a low enough
level of aggregation is a serious obstacle. Furthermore, as is noted below, much of the R&D
spending occurs outside the firm or industry in
question.

None of these approaches seem suitable for
modeling the development of new processes in the
engineering sense. What makes new processes
especially difficult to deal with is that they
often require a considerable effort in basic
science, as well as in the overcoming of engineering problems in order to implement a process
for which the chemistry and physics are well
understood. Basic science connotes an extremely
risky activity in which payoffs are subject to
high variance. Rosenberg [28], for example,
points out that greatly improved knowledge of
medicine took many centuries to develop, despite
the continuous flow of resources devoted to
medical problems during that period. Detailed
knowledge of a product will supply some a priori
ideas about the portion of input space which could
be fruitfully investigated, but this degree of
detail may be more than is desirable for purposes
of general economic modeling.

For several of the innovation types discussed above it is rather straightforward to determine the impacts of relative and absolute price
changes. Proportionate increases in input prices,

26

for example, would increase the payoff to input-reducing innovation in general. Relative changes in input prices would favor innovations permitting substitution against the relatively higher priced inputs and also stimulate the search for new processes which economize on those inputs. Implications of this nature follow from the models discussed earlier in this section. The effect of price changes on capacity-increasing innovations is more difficult to analyze: an increase in capacity leads to a lower average cost with constant input prices, but this will be offset to some extent by the higher costs associated with a price increase in the nonlinear inputs which constrained capacity - generally plant and equipment. Thus, it is not clear what effect a rise in the price of capital goods will have on capacity, when increases in capacity require additional capital goods; innovations which reduce capital intensity may become more attractive than those which increase capacity.

The effects of environmental and workplace regulations have not been studied formally within our innovation model. Nevertheless, a few tentative generalizations might be offered. Some processes may be so harmful to the environment or worker health and safety that the development of a new process is necessary if continued production of the product is to take place. In other cases it may be necessary to substitute against an input which has deleterious effects on the environment or worker health and safety; for example, innovations designed to permit the use of fuels other than coal might be generated. Note that a change in fuels may not be a simple substitution: chemical process technology is highly integrated in the sense that the same hot air or steam is used for different purposes as it cools; an attempt to change any part of the process may require changes throughout. For example, in steam reforming of ammonia, the same input is used as both fuel and feedstock, with purge gases being used as supplementary fuel. Furthermore, as we shall see in Chapter 5, process modifications for control of regulated effluents may increase plant fixed and operating costs in various ways and may thus stimulate a search for factor-saving innovations in the absence of factor price changes. Finally, capacity-increasing innovations may be inhibited if an environmental regulation takes the form of limits on total pollutant emissions at a

27

given site. Of course, this effect may be partially offset by reductions in input use or by substitutions against the offending input.

2.5   AN EMPIRICAL MODEL FOR PROCESS INNOVATION

The process innovation model developed in this chapter yields testable hypotheses concerning changes in input coefficients and in maximum plant capacity. Specifically, we hypothesize a relationship between innovation, as measured by movements in input coefficients for a given process on the one hand, and factor price changes on the other. We test this hypothesis with data for the steam reforming process and report on results in Chapter 6. Our first step is to establish that observed coefficient changes were innovations rather than substitution. We then explain the time series of input coefficients with input price data in order to examine the extent to which the change is in the direction consistent with a theory of process change.

A second empirical study is concerned with hypotheses about changes in upper limits to capacity. As noted above, efforts to increase these limits are hypothesized to depend upon the price of ammonia, transportation costs, and other variables. These tests are also reported in Chapter 6.

Finally, it has not been feasible to include changes in environmental, health, and safety regulations as driving forces for innovation in our modeling work. (See Chapter 5 for explanation of why these forces have been excluded, and Section 6.3 for a discussion of how they might be included in similar studies.)   Regulation of natural gas prices is captured through the price mechanism in the model. Problems with gas availability in recent years may have stimulated innovative activity on processes or inhibited selection of sites on interstate pipelines. We believe that it is too soon to capture such responses empirically for the purposes of our modeling effort.

3. A History of Regulation of the Nitrogen Fertilizer Manufacturing Industry

3.1 INTRODUCTION

The sophisticated chemical processes used for the manufacture of nitrogen fertilizers and the caustic and explosive characteristics of some of the products involved threaten the health and safety of workers and the public and raise the possibility of air and water pollution. As a result, this industry, like many others, has been the subject of workplace and public health and safety regulation for a considerable period and, more recently, the subject of environmental regulations. This chapter focuses on the history of these regulations.

More specifically, the chapter begins with a broad view of the general types of regulations that have affected the industry and explains our reasons for focusing on two general categories of regulations: workplace health and safety and environmental protection. We then develop the regulations designed to provide worker health and safety, including a review of the Occupational Safety and Health Act of 1970, the Walsh-Healey Public Contracts Act, state boiler safety laws, and state workmen's compensation laws. [1] Next we consider environmental regulations, focusing on the Federal Water Pollution Control Act Amendments and the Clean Air Act Amendments of 1970. [2] Our goal in this chapter is to describe the regulations which are relevant to the nitrogen fertilizer manufacturing industry and establish the date upon which they began to have an impact on the industry. We first describe the regulations as they existed at the time of the research, up to December 1976, and then proceed backwards in time through their predecessors. We also introduce various implications of these regulations for ammonia process technology. These concepts are developed further in Chapters 5 and 7.

"Regulation," a nebulous term, encompasses a range of impositions on a person's or firm's conduct. In the ordinary sense we think of regulations as those requirements or conditions imposed for engaging in certain conduct. For example, prohibiting or limiting effluent emissions is a regulation, as are those measures requiring businesses to maintain specified safety standards or requiring licenses to engage in specified conduct. But the concept goes much further. Arranging the tax structure to encourage specified business decisions -- such as increasing capital investment, regulating prices to make one factor of production (such as interstate natural gas) more desirable than its alternative (intrastate natural gas), and curtailing the amount of a resource (such as farmland through land bank programs) thereby enhancing the value of high productivity for the resource use -- are all forms of regulation.

For purposes of this study, we distinguish between those regulations that impose constraints directly upon nitrogen fertilizer manufacturing processes and those that indirectly affect the processes by altering input factor prices or the demand for the product. We focus on the direct regulations because their effect on processes may be more visible. It is true that, in given circumstances, the indirect regulations may be as important as the direct. For example, as natural gas resources become exhausted, stringent allocation of natural gas may be imposed, and the industry may no longer have reliable sources for gas. We believe that the distinction can be made for this study since its goal is not to demonstrate exhaustively how every conceivable regulation may have an impact on the industry. Further studies can be undertaken to consider the effect of any other complex of regulations. In any case, it does seem necessary to understand the regulations which impose constraints directly upon the process in order to understand other regulations.

Our study indicates that the regulations that directly affect the industry's processes fit generally into the categories of workplace health and safety regulation and environmental regulation. The indirect regulations include natural gas, transportation, general business regulation,

tariffs, farm regulation and assistance, Tennessee
Valley Authority activity, tax laws, and others.
In the following section, we review briefly these
indirect regulations.

3.2   INDIRECT REGULATION OF AMMONIA PRODUCTION

Natural gas regulation has become a matter
of major importance to the industry since it has
been widely predicted that there will soon be
supply shortages. The impact of that regulation
is reflected in the industry through price and
availability of natural gas. As of 1977, gas was
available, and it appeared likely to remain so in
the near future since Congress had made the
fertilizer industry one of the highest priority
purchasers. [3] Limited capacity on interstate
pipelines, however, had led many private distribution firms to refuse to supply natural gas to
new customers. Therefore, the industry faced some
availability difficulties. Fertilizer industry
consumers who chose to locate plants at or near
the well-head avoided this problem but confronted
another. Under the then existing system of
federal regulation of natural gas, the price of
gas sold interstate is set under the Federal Power
Act [4] while gas sold to intrastate industrial
consumers was largely unregulated. As of 1977
and for the forseeable future (unless federal rate
regulation of natural gas is terminated), the
market price for intrastate sales was appreciably
higher than the price set by the FPC for interstate sales. Thus, the consumer who located near
the well-head to avoid the availability problem
was required to pay a higher rate for the gas
consumed. Finally, shortages in natural gas
supplies for home heating may produce legislation
giving the federal government the power to allocate intrastate natural gas supplies, which would
result in uncertainty in the availability of
natural gas even when purchased at the well-head.
Consideration has been given to such allocations,
at least among interstate pipeline firms. [5]
For example, on February 21, 1977, Senator Talmadge
proposed an amendment to the National Gas Act "to
provide a priority system for certain agricultural
uses of natural gas" including use as feedstock
and fuel in the production of fertilizer.
Transportation regulation has also had an
important impact on this industry, but, again, its

impact is felt in terms of the availability and
relative prices of the various modes of transportation. For example, since the Texas City
disaster involving the massive explosion of
ammonium nitrate, the ports in this nation have
been closed to that product. [6] Therefore,
ammonium nitrate can only be transported by long
distance rail and by truck within smaller markets.
Barge transportation, which has proved effective
for ammonia and urea, is unavailable for ammonium
nitrate. Likewise the Department of Transportation
and the Interstate Commerce Commission regulate
the hauling of the products both with respect to
general safety concerns and to the specific concerns of ammonia transportation, given its particularly caustic and high pressure characteristics. [7]

A host of other regulations have indirect
implications for the industry. Tariffs on international food and fertilizer trade and American
foreign aid policies may affect the demand for
fertilizer. [8] Regulation of the structure of
industries through the antitrust laws [9] and, more
remotely through the Federal Power Act [10] and the
Public Utilities Holding Company Act [11] may
affect the organization of the fertilizer industry
as well as the prices charged by natural gas and
other input suppliers. Laws administered by the
Securities and Exchange Commission, [12] the
Federal Reserve Board, [13] and others may affect
the availability of financing for the industry. [14]
Demands for fertilizers may have been affected
by state and federal funding of Land Grant
Colleges, [15] Agricultural Extension Services,
and County Agents, all of which have contributed
to the sophistication of American farming. [16]
Federal land bank programs have also enhanced
demand by encouraging high productivity per
acre. [17] Federal parity policies for farm
prices affect demand by assuring the farmer a
minimum price for his commodity, which if high
enough may justify the farmer's using more
fertilizer to marginally increase yield. [18]

The Tennessee Valley Authority (TVA) has had
a "regulatory" effect upon the industry. (See
Section 1.1.) TVA has produced some innovations
in fertilizer blending, has spearheaded the
development of some higher analysis fertilizers
(although more in the phosphate part of the
industry than in the nitrogen), and has undertaken
demonstration projects such as a coal gasifier

32

to provide part of the feed to its Muscle Shoals, Alabama, ammonia plant. Its activity appears to have provided more information for the private sector, to have stimulated the private sector to intensify its own research and development, and to have created a competitive situation which forced some inefficient fertilizer manufacturers out of the market. [19] For years, the TVA has maintained elaborate statistics on its and private industry's activity. [20] This appears to have helped shape an industry in which competitive success depends on efficiency and scale and not on trade secrets and patents.

Federal tax laws also appear to have had an effect on the industry. Certainly, investment tax credits enhance the economic feasibility of constructing new manufacturing facilities, and those have been available off and on over the years. [21] The fact that this is a capital intensive industry gives the credit special significance. Further, the special treatment given cooperatives under the Internal Revenue Code also seems to have been significant for the structure of the industry. [22] As already indicated, manufacture of nitrogen fertilizers is a very refined process requiring high efficiency. Expert personnel must be available but are necessary only on an occasional basis. Smaller firms, having considerable difficulty supporting such expert staffs, may find it impossible to compete in the industry unless they have some other competitive advantage. It has been suggested to us that the tax breaks given cooperatives provide that other advantage, and partially explain the presence in the industry of some smaller fertilizer producers organized as cooperatives.

Finally, no review of regulations affecting this industry can overlook those regulations which affect all firms. These involve compliance with Social Security and Federal Income Tax rules, recordkeeping to ensure equal opportunity employment practices, and many others.

With this background, we can now turn to those regulations that more directly address nitrogen fertilizer manufacture.

3.3 WORKPLACE HEALTH AND SAFETY REGULATION

At a basic level, workplace health and safety considerations are a necessary part of the manufacturer's calculus. Accidents that involve danger to employees necessarily also involve interruption of production. Consequently, in many instances, the responsibility that is imposed by a new regulation is not greater than that already assumed by some employers. The introduction of workplace health and safety regulation, however, apparently is a response to the perceived need to impose on the employer a greater responsibility in this area than he automatically assumes because of economic considerations.

As of 1977, the most pervasive workplace regulations were those promulgated under the Occupational Safety and Health Act of 1970 (OSHA). These regulations apply to all firms engaged in interstate commerce including all nitrogen fertilizer manufacturers. Between 1971 and 1973, OSHA promulgated a series of regulations, some of which were under review and revision in 1976. Before enactment of OSHA, the federal government's role in regulating workplace health and safety had been confined to overseeing workplace conditions of firms under contract to the federal government. That activity started in 1936 with enactment of the Walsh-Healey Public Contracts Act and appears to have become a more intensive, or at least more formal, activity in 1960.

Regulation on the state level has a much longer history. Boiler safety regulations began appearing in the 1910s although some states deferred enacting these regulations twenty or more years thereafter. As of 1977, state boiler regulation was common. The purpose of OSHA and boiler safety laws is to ensure, to the extent feasible, that workers will not be injured or become ill as a result of their employment. Another body of law, workmen's compensation laws, primarily serves the alternative purpose of providing an injured or ill worker with compensation for the disability suffered. Many such laws were enacted between 1910 and 1920, and many that originally excluded "occupational diseases" from coverage of their laws were modified between the late 1930s and early 1950s to allow compensation for such diseases. Prior to enactment of workmen's compensation laws, the workplace was regulated by the common law of torts: an employee would sue

34

to recover damages for the amount of the injury
suffered, but only if (1) the employee had not
"assumed the risk" of the accident, (2) the
employee did not contribute to the cause of the
accident by his or her own negligence, no matter
how slight, and (3) the injury was not caused by
a "fellow servant," that is, fellow employee, of
the employer. Often these were impossible
limitations to surmount.

It is the constellation of all these laws and
regulations that constitutes workplace health and
safety regulation today for the nitrogen fertilizer
manufacturing industry and all others. These are
the regulations which will be discussed in this
part.

The Occupational Safety and Health Act (OSHA)

From the point of view of the nitrogen
fertilizer industry, the most significant workplace regulation is OSHA. This section will
outline the OSHA regulations affecting fertilizer
manufacture in 1977, discuss two related rulemaking proceedings then underway, and analyze the
practical implications of OSHA for the industry.

The OSHA Regulations. The OSHA regulations
operate in two ways: First, there are a series
of regulations promulgated by the Department of
Labor that set standards for maintaining worker
health and safety, and second, there is the
"General Duty Clause," which requires that the
employer keep his places of employment free from
"recognized hazards that are causing or are likely
to cause death or serious physical harm to his
employees . . . ." [23] Two of the regulations
promulgated under OSHA, both of which were being
revised in 1977, deal specifically with nitrogen
fertilizer manufacture: the regulation of storage
and handling of anhydrous ammonia, effective
February 15, 1972 (the "storage and handling" regulation) [24] and the regulation of exposure of
workers to gaseous ammonia, effective June 27, 1974
(the "worker exposure" regulation). [25] In
addition, a series of other OSHA regulations apply
to this manufacturing industry. For example, centrifugal compressors used in the more modern manufacturing facilities are noisy and thus must be
evaluated in light of the OSHA noise exposure regulation. [26] All manufacturing facilities have
ladders, scaffolds, and other work surfaces which are

subject to specific OSHA regulations. [27] Where
there is no specific standard, the General Duty
Clause controls. Thus, either workplace health
and safety hazards are the subject of explicit
regulations, or the employer is under a general
duty to protect his employees from hazards that
are recognized and serious.

The Worker Exposure Regulation. Originally,
OSHA was authorized to adopt regulations which
set workplace health and safety standards at the
level that had previously been adopted by other
federal regulatory authorities or that had been
set by private standard setting organizations
whose activities had a national consensus. [28]
The regulation promulgated under the Walsh-Healey
Public Contracts Act [29] was the predecessor for
OSHA's worker exposure regulation. That standard
and the present OSHA standard prohibit employers
from exposing their employees to concentrations
of ammonia greater than 50 ppm on an 8-hour
time-weighted average. [30] The currently proposed modification would prohibit exposure at
any time to a concentration of ammonia of more
than 50 ppm based upon a sample taken for no more
than five minutes. [31] The effect is to set a
50-ppm ceiling rather than a 50-ppm average
exposure, thus eliminating the possibility of
occasional exposures at significantly higher
levels.

OSHA's public statements on the proposal
explain its view that there is "no evidence of
chronic illness from a long-term exposure" to
ammonia. [32] Therefore, they conclude the
weighted average is inappropriate. The concern
that the regulation addresses is the immediate
effect of exposure to a significant amount of
ammonia. For that, they argue, a ceiling is
appropriate. Some industry members strenuously
object to the proposed ceiling; they argue that
compliance would be too expensive, that restriction
of exposure on a permanent basis is close to
technically impossible, that the regulation is
unnecessary since workers will smell concentrations
of ammonia and leave the area before it becomes
dangerous, that medical examination, equipment
and recordkeeping requirements are cumbersome and
unnecessary, and that restricting to that level
of exposure is not necessary to protect worker
health and safety. [33] One firm argues that
there is evidence to support the view that

36

employees regularly working in areas exposed to
ammonia become accustomed to it and can tolerate
even higher levels than others. [34]

In its summary of the research in the area,
OSHA appears to tacitly admit Allied Chemical's
assertion that the "proposed standard is based on
a comfort level for all exposed employees
regardless of the frequency or the extent of their
exposure." [35] Research tends to suggest that
50 ppm is approximately the level at which the
odor of ammonia becomes noticeably offensive and
causes some minor irritation but does not present
any major short-term or chronic health problems. [36] If this is a "comfort level" regulation and if the department does have the
authority to promulgate it, then it would seem
that OSHA's standard-making authority is based
upon a very liberal interpretation of "Safety and
Health." This, in turn, may suggest that by
proposing this regulation, OSHA was testing its
power to extend its jurisdiction beyond points
it had already travelled.

The public comment period on these proposed
regulations ended April 26, 1976 [37] and, as
of the end of our research in June 1977, promulgation of a final regulation awaited only
delivery of an economic impact study being completed for OSHA by a private consultant and a
decision by the department. [38] Because of the
"comfort level" aspect of the regulation, the
outcome may prove to be quite important for the
development of OSHA's authority.

The Storage and Handling Regulation. Within
OSHA, there is an administrative separation of
"health" and "safety" standards development. [39]
The worker exposure regulation is a product of
the health standards setting; the storage and
handling regulation is oriented to safety. The
predecessor for the latter regulation is a
standard promulgated as a "national consensus"
standard by the American National Standards
Institute (ANSI), [40] although the industry
believes that ANSI's standard did not in fact
represent a national consensus. For reasons not
entirely clear, ammonia manufacturers were
exempted from the ANSI standard and are therefore
exempted from the OSHA storage and handling
regulations. The regulation applies to some of
the people involved in the transportation,
blending, distribution, and use of anhydrous

37

ammonia. Further, by the terms of the worker exposure regulation, people subject to the storage and handling regulation are not subject to the worker exposure regulation.

As previously noted, the Department of Labor initiated a proceeding which may result in modification of the regulation. [41] The department has announced that it is aware of criticism of the existing standards, has summarized some of those criticisms, and has solicited suggestions on how the regulation should be modified. It had not in 1977 published a proposed amendment of the regulation. The department's announcement, however, does raise the questions of whether ammonia manufacturers should be exempted from the storage and handling regulation and whether those subject to this regulation should also be subject to the worker exposure regulation.

Prior to the department's announcement of the revision proceedings for the storage and handling regulation, representatives of The Fertilizer Institute (TFI), a trade association of fertilizer manufacturers and distributors, had pressed for its revision. Their primary argument had been that the regulation was out-of-date because ANSI had revised its standard three times since the OSHA adoption and because the more current versions of the ANSI standard had been developed in cooperation with TFI. Part of the purpose of this rulemaking proceeding was clearly to consider adopting the ANSI revisions. The department had gone beyond that suggestion, however. It had sought discussion of whether particular provisions of the regulation should be changed to "require a level of performance which would ensure employee safety without specific means of meeting that performance level." [42] As an example of the reasons for this possible change, the department noted:

> the requirement in §1910.11(h)(2) that 'shell or head thickness of any container shall not be less than three sixteenths inch' has been criticized as restricting the development of thinner but equally safe containers. [43]

In its section-by-section description of criticisms of the regulation, the department stated that it had been suggested that the thickness requirement be deleted. [44] This suggestion is not necessarily a

38

satisfactory response to the suggestion of substituting performance standards for specified design requirements. In order to develop a performance standard it would be necessary to ascertain which properties of the shell provide the margin of safety achieved by the thickness requirement and to state those properties as performance requirements. Such a performance regulation clearly creates a greater incentive for innovation in containers since it would allow the possibility of thinner container walls. However, the performance standards must be developed to assure that a thinner-walled vessel provides the required margin of safety while assuring that the standards are not so tight that they, in effect, actually require the three sixteenths inch thickness. Further, the standards must be designed in such a way that it is practical for OSHA enforcement people to ascertain whether the standards have been achieved when the wall is thinner than presently required. To overcome this latter problem, the department alluded to the possibility that the regulation will be written with performance standards and accompanied with guidelines that show in design terms how those standards can be achieved. [45]

Other Applicable OSHA Regulations. Substantial portions of the remaining OSHA regulations apply to the manufacture of nitrogen fertilizer. These other regulations can be divided into two categories, those that are particularly relevant because of the peculiar nature of the fertilizer manufacturing process and those that are relevant simply because they would apply to any complex manufacturing process. The former include regulations regarding emergency exits, ventilation, noise, nitrous oxide (in the case of nitric acid production), pressure vessel and storage tank integrity, explosives (in the case of ammonium nitrate production), eye and face protection, respiratory protection, and air contaminants in addition to ammonia (such as nickel particulates). [46] The latter include regulations regarding walking and working surfaces in plants, handling of compressed gas generally, handling hydrogen and, in the partial oxidation process, oxygen, storage of liquified petroleum gases, compressed gas cylinders, safety relief devices for storage and cargo, compressed gas tanks, machinery and machinery guarding, welding, and so forth. [47]

39

These regulations can further be divided into those that may directly affect the technological design of a production process and those that represent an increased capital cost for the process. For example, a restriction that limits the permissible noise levels in the workplace might, all other things being equal (and in this case they are not), tip the balance in favor of a reciprocating compressor instead of a louder centrifugal compressor. On the other hand, a regulation that requires a large number of fire exits may merely increase the cost of constructing a plant because the requirement may be more than otherwise required by an architectural standard, building code, or local fire safety authority.

Our study indicates that these other OSHA regulations have not led the nitrogen fertilizer industry to require design modifications involving the substitution of one production process for another. Thus, even though the regulations may have affected the cost of manufacturing nitrogen fertilizer, the effect is in the nature of increasing capital or operating costs, not forcing new technologies. Further, as far as we can tell, the impact of these regulations has been slight relative to the industry's overall operations.

Compliance with OSHA Regulations. Compliance with OSHA regulations is enforced by a combination of activities of several groups, including official enforcers (either employed by OSHA or state agencies in those states that have taken over primary responsibility for OSHA enforcement), employees, and inspectors employed by insurance companies. Here we discuss the role each plays and the interrelations among those roles.

One of the most startling features of OSHA is the scope of its jurisdiction. In Section 2(b) of the Act, Congress states:

The Congress declares it to be its purpose and policy, through the exercise of its power to regulate commerce among the several states and with foreign nations and to provide for the general welfare, to assure so far as possible every working man and woman in the Nation safe and healthful working conditions . . . . [48]

Under the section entitled "Applicability of this Act," the Act provides that it "shall apply with

respect to employment performed in a workplace in
a State, the District of Columbia," and every
other possession and territory of the United
States. [49] Thus, the scope of the Act extends
to virtually every employer in the United States.
Based upon that broad jurisdiction and the
relatively limited resources devoted to OSHA, one
has a tendency to assume that there is very little
chance that any given employer will be punished
if he ignores the OSHA standards. [50] This in
turn suggests the hypothesis that OSHA does not
effectively represent a regulatory imposition on
any employers except those specifically targeted
for review by the department. If that hypothesis
were borne out by the facts, we would have
identified an area in which a stated regulation
is meaningless. Therefore, we have delved into
the question rather carefully and have concluded
that although the regulation is clearly different
from the way it appears on its face, because of the
enforcement pattern, it is not meaningless. [51]
An outline of the enforcement provisions of
OSHA is necessary to understand the incentives
for compliance by an employer. When an inspector
discovers a violation at the workplace, he is
obligated to issue a citation for the violation. [52] If a citation is issued, the
department must soon thereafter issue a notice of
the penalty that will be assessed upon the
citation. [53] And, if the department "has reason
to believe that an employer has failed to correct
a violation for which a citation has been issued"
within the time specified for correction, the
department is required to issue a notice of a
further violation and assess an additional
penalty. [54] Penalties for serious violations
must be assessed; the department has the discretion
to refuse to assess a penalty for nonserious
violations. [55] The penalty runs up to $1000
for each violation and each day of failure to
correct the violating condition. [56] An employer
who "willfully or repeatedly" violates the OSHA
standards is subject to a penalty up to
$10,000. [57] OSHA's pattern has been to assess
very small fines, on the order of $25 to $100,
for first citations. There is, of course, the
usual complement of procedures for substantiating
the inspector's violation finding [58] and for
administrative and judicial review of the department's ruling. [59]

This penalty mechanism, though, is irrelevant unless an inspector has discovered the violation. In view of the limited number of inspectors for an enormous number of employers covered by OSHA, it is unlikely that any given employer will be inspected and cited unless OSHA is specifically notified of a problem at a workplace. Further, even where an inspector appears on the premises, he may not be able effectively to inspect for health standard violations. Inspections for such violations require technical sensing equipment that inspectors may not have available at the inspection site, time for repeated tests that the inspector may not have because of other inspection responsibilities, and technical sophistication that not all inspectors have. OSHA administrators seem aware of this problem and are apparently trying to upgrade the quality of the inspection for health standard, as well as safety standard violations. [60] In one respect this may not be a particular problem for the key health standard in nitrogen fertilizer manufacture -- worker exposure to ammonia fumes. Since such fumes are quite noxious, the inspector's sense of smell will at least alert him to the need to set up sensing equipment.

The act provides for bringing information of threatened or actual violations to OSHA's attention even when a site has not been inspected. Under section 8(c) of the Act, [61] the Departments of Labor and Health, Education, and Welfare are authorized to require employers to make records and reports of work-related deaths and injuries or illnesses, unless they are minor. The department has issued regulations implementing those recordkeeping and reporting requirements. [62]

Information regarding possible violations may also come from employees. Employees or their representatives are specifically authorized to bring hazardous conditions to the attention of the department, and if the department determines there are reasonable grounds to believe that a violation threatening physical harm exists or that there is an imminent danger, a special inspection must be undertaken. [63] Further, no employee may be discharged or discriminated against on the grounds that he complained of a violation under the Act. This provision is enforceable by permitting OSHA to force a violator to reinstate and provide back pay to a wronged employee. [64] The usefulness of information provided by employees is enhanced

42

by the fact that the Act and regulations under it,
in many instances, stipulate that information
regarding requirements imposed on employers and
violations by employers be explained to their
employees. [65]

In addition to enforcement by OSHA, there is
another mechanism which encourages implementation
of OSHA standards. It arises from the response
of employers' insurers to the Act. Many OSHA
standards, though certainly not all, provide
important assistance to the insurer who wishes to
see risks, and therefore loss claims, minimized.
As a result, many, if not all, private insurance
underwriters require that their inspectors examine
the condition of an employer's workplace in light
of OSHA standards in order to determine whether to
underwrite the insurance risk. Significant noncompliance with OSHA standards then puts the
employer at risk of being unable to obtain insurance, in addition to the risk of OSHA enforcement
action.

No doubt an insurance company may underwrite
an insurance policy even if the employer is not in
full compliance with OSHA, if only because there
can be honest disagreements about the value of
any given OSHA standard. But even in that case the
insurer's inspector has an incentive to inform the
employer that there is noncompliance with OSHA
standards in some particular area. If the inspector does not and a subsequent OSHA inspection
results in a citation, the employer, dissatisfied
with the information he obtained from the
insurance inspection, may direct his insurance
business elsewhere. As a result, insurance
companies offer these inspection services as part
of their service. In many instances the mere
fact that an employer knows that he is not in full
compliance will stimulate corrective action.

All of this suggests that the enforcement
of OSHA is accomplished through a multi-tiered
program that enhances the prospects for achieving
worker health and safety while reducing the
likelihood that expensive, ineffective requirements
will be imposed on employers. At one level there
is the enforcement machinery instituted by OSHA
itself: inspections and the issuance of citations.
Yet the infrequency of inspections because of
manpower limitations, coupled with the Department
of Labor's discretion in applying sanctions and
conducting inspections, militate against rigid
application of the law. Organized labor and

individual employees may, however, overcome some
of the limitations in the inspection machinery
through their power to force OSHA to initiate
inspections of particular employers upon a showing
of good cause. In the meantime, private insurers
conduct frequent inspections of all employers.
If an employer maintains conditions that are
seriously out of line with OSHA standards, the
insurers will refuse insurance. Other aspects of
noncompliance found by insurance inspectors are
likely to be reported to the employer, who is then
left with the decision of complying or risking a
citation from an OSHA inspection.

Certainly, there are difficulties with this
situation that make it less than ideal. For
example, if a particular OSHA standard is patently
absurd, insurers may ignore it. Further,
employers will be reluctant to refuse to comply
with a standard even though they may consider it
absurd and recognize the possibility that they
will never be called to task for noncompliance.
Also employees may fear employer reaction to
their filing complaints even though the employer
is restricted from discharging or discriminating
against complaining employees. Finally, there is
always the possibility that the employer remains
unconvinced that a given standard enhances health
or safety. But without further study it cannot
be said that this system is inferior to such
alternatives as an inflexible system of enforcement
that requires absolute compliance in every case or
a system with no enforcement capability at all.

OSHA Beyond 1976. The impact of OSHA on the
nitrogen fertilizer industry through 1976 has
been mixed. The anhydrous ammonia storage and
handling regulation had not been applied to the
industry. The ammonia exposure standard and the
decibel limit regulations were federally
established regulations, and impressions derived
from the industry indicate that these are fairly
consistent with standards maintained before OSHA.
On the other hand, the introduction of the OSHA
enforcement mechanism may have stimulated compliance efforts.

How standards will be modified in the future
is not clear. It is clear, however, that OSHA
has embarked on developing its own standards to
replace those ·adopted from other sources in the
1971-73 period. This development may create new
requirements that very seriously affect the

economics of the manufacturing process. And, with
the Carter Administration currently indicating
broad changes in the focus of OSHA standard
setting and enforcement, the longer run implications for process economics are even more
uncertain.

On the legislative side, various bills have
been introduced to amend the Act. One that was
particularly significant in light of OSHA's
enforcement structure was introduced by former
Ohio Senator Robert Taft. It would have created
a separate staff in OSHA to consult with employers
about compliance. Except where the OSHA consultant found serious violations, he would not be
permitted under the bill to inform the enforcement
arm of OSHA of his findings. It is not clear
what impact this would have on the enforcement
structure we outlined above. [66] It was, however,
the absence of OSHA consulting services which
encouraged the insurance industry to get into
that business. [67] If OSHA consultants are
available, private industry may be less interested
in providing the service. Further, a major
deterioration of the law would occur if an employer
could defend against a citation on the grounds
that he had asked for compliance advice from the
department and had not yet received that advice
when he was inspected and cited. Finally, if
adequate advice is available from the insurance
industry, one wonders whether additional federal
resources are better devoted to an improved
enforcement capability at OSHA than to providing
services already available in the private sector.

The Walsh-Healey Public Contracts Act

One of the first federal excursions into the
regulation of workplace health and safety occurred
with the enactment of the Walsh-Healey Public
Contracts Act. [68] In this subpart, we will
describe the coverage of the Act, the development
of standards under it, and its enforcement and
significance.

Standards. Under the Walsh-Healey Public
Contracts Act, Congress established a series of
regulations for private firms which were under
contract to the United States "for the manufacture or furnishing of materials, supplies,
articles, and equipment in any amount exceeding
$10,000 . . . ." [69] Such firms were required

to include in their contracts with the United
States provisions which assured minimum wages,
40-hour regular working weeks, minimum ages for
employees, and

> That no part of such contract will be per-
> formed nor will any of the materials,
> supplies, articles, or equipment to be
> manufactured or furnished under said con-
> tract be manufactured or fabricated in any
> plants, factories, buildings, or surroundings
> or under working conditions which are
> unsanitary or hazardous or dangerous to the
> health and safety of employees engaged in
> the performance of said contract. [70]

The Act excludes from coverage purchase of
"materials, supplies, articles, or equipment as
may usually be bought in the open market." [71]
On its face this might suggest that nitrogen
fertilizers are excluded from coverage. However,
a 1957 United States Court of Appeals decision,
in a case concerned with the Act's applicability
to purchases in the bituminous coal industry,
suggests otherwise. [72] The Court adopted the
Secretary of Labor's argument that the exclusion
from coverage only applies to purchases "the
Government itself is authorized to make in the
open market," [73] saying

> The Secretary's interpretation ... has
> persisted throughout the administration
> of the Act and has been the basis upon
> which appropriations have been made. [74]

Thus it appears that contractual purchases of
nitrogen fertilizers would extend the coverage of
the Act to the manufacturer, even though the
fertilizer might simultaneously have been purchased
in an open market transaction from a supplier.
To our knowledge there has been only a
minimum of federal contracting with the nitrogen
fertilizer industry, that being for the occasional
purchase of explosive grade ammonium nitrate,
especially in times of war, [75] and the occasional
purchases of fertilizer in connection with agricultural assistance programs. [76] Certainly, the
presence of a workplace regulation for some
industries can have a "leading" effect for setting
all industries' standards on a given point and, to
that extent, the Walsh-Healey regulations may also

have indirectly affected the nitrogen fertilizer industry. In any event, the Act takes on special significance because under it were promulgated regulations that limited the permissible worker exposure to ammonia. These regulations were the "Federally established standards" which were adopted as an OSHA regulation after its enactment.

For the first twenty-four years after the law was enacted, the Labor Department did not set any regulations regarding prohibited working conditions. This can be explained partially by the fact that workplace health and safety concerns were clearly secondary to the minimum wage and other wage regulations of the bill. [77] Another explanation can be found in a caveat to the section prohibiting unhealthful and unsafe working conditions:

> Compliance with the safety, sanitary, and factory inspection laws of the State in which the work or part thereof is to be performed shall be prima facie evidence of compliance with this subsection. [78]

One fair reading of that provision is that the Labor Department could not establish any work rules which were stricter than those established by the state in which the work is being done.

In the last days of the Eisenhower Administration, however, the department issued a statement which took the view that the provision did not prohibit the department from establishing a "uniform national standard." [79] Rather, the department concluded, it was appropriate for it to publish a general statement regarding what it believed health and safety conditions ought to be. It recognized that there might be debate on whether some standards were more rigorous than would be necessary to comply with the Walsh-Healey workplace health and safety requirements. Therefore, it indicated that it would hear evidence in adjudications that a standard is excessive but indicated that a person claiming that a standard is too rigorous would have to present his case for that view. [80] Since the department took the view that the "regulations are rules of agency procedure or practice delineating facts which will be officially noticed in enforcement proceedings," [81] it promulgated the regulations without advance notice or opportunity for comment. We have not found any indication that the propriety

of these regulations or their promulgation was
formally challenged.

Included in the standard was a maximum for
worker exposure to ammonia of 100 ppm on an 8-hour
weighted-average. The standard further provided
that a worker could not be exposed to any amount
more than the maximum or exposed to more than one
of the various substances (including ammonia) that
were the subject of the regulation, unless given
prior approval by an industrial hygienist who was
informed on the particulars of the additional
exposure. [82] This latter provision is quite
similar to the proposed restriction of exposure to
ammonia to 50 ppm based upon a five-minute sample
currently under consideration by OSHA (see Section
3.3), unless there is an implicit assumption that
the hygienist would automatically approve moderate
but substantial additional exposure in the event
of unanticipated leaks and the like.

A rulemaking proceeding was undertaken in
1963 to revise the 1960 Walsh-Healey regulations, [83] but the process seems to have been
dropped. The next change in the regulations, as
they related to the nitrogen fertilizer industry,
did not occur until 1969. On September 20, 1968,
the Secretary of Labor published a proposed
revision of the regulations, the need for which,
he said, was based upon the department's experience
and requests by interested groups. [84] The proposed new standard for ammonia exposure was
50 ppm on the 8-hour weighted-average. [85] Noting
that the

> threshold limit values [which included the
> 50 ppm standard for ammonia] represent
> conditions to which employees may generally
> be exposed repeatedly for long periods
> without harm, [86]

the prohibition against substantial episodic
exposure without an industrial hygienist's consent
was eliminated for all but those episodes which
were so extreme that zero exposure for the remainder of the day could not reduce the average
to 50 ppm.

In the last days of the Johnson Administration,
Secretary Willard Wirtz published final rules
pursuant to the proceeding initiated with the
September 20, 1968 announcement. [87] The exposure regulation in its final form was changed by
eliminating the table of exposure standards and

inserting a prohibition of exposure to amounts
greater than those specified in the "Threshold
Limit Values of Airborne Contaminants for 1968" of
the American Conference of Governmental Industrial
Hygienists, except with respect to certain
chemicals other than ammonia. [88] The Secretary
of Labor succeeding to office after January 20,
1969, George P. Shultz, postponed the effective
date for the new rules which had been set for
February 16, 1969. [89] Then, on May 20, 1969,
the department published and adopted a new version
of the regulations. The standard for ammonia
exposure in these regulations is apparently the
same as the analogous standard in the final
Secretary Wirtz version [90], although the wording
is not identical. [91] This regulation has continued in effect to the present. [92]

     Enforcement and Compliance. While the Act
provides for damages to be paid to the United
States for violation of the wage regulations, the
only remedy the government has for workplace
health and safety violations is to cancel a contract and procure completion of the contract
elsewhere, with any additional costs involved
therein to be borne by the violating contractor. [93] The limited scope and the severity of
this mode of enforcement suggests that formal
enforcement of the Act would be limited to only
extreme cases. This was confirmed by the Study
Group on Disease and Injury on the Job, organized
through the auspices of Ralph Nader. [94] For
example, they summarized:

     In Fiscal 1969, safety engineers in the
     regional offices of the [Bureau of Labor
     Standards of the Department of Labor]
     performed safety and health inspections
     in only 5 percent of the firms covered
     by Walsh-Healey. Out of a total of 75,000
     firms affected by the Act, only 2,929 were
     inspected. In 95 percent of these estab-
     lishments, inspectors found safety
     violations -- indeed, a grand total of
     33,378 violations. Out of this number,
     only 34 formal complaints were issues, 32
     hearings were held, and 2 firms received
     the ultimate sanction of blacklisting
     [for future government contracts for up
     to 3 years]. In Fiscal 1968, the [Bureau]

inspected 1,570 firms, found 48,646 violations, issued 28 complaints, and blacklisted 3 firms. [95]

This is to be compared with current OSHA enforcement rules that make it possible for many offenses for fines to be on the order of $25 for each violation and that require the OSHA inspector to issue a citation for every violation he observes. [96]

Perspectives on the Significance of the Walsh-Healey Act. It is difficult to ascertain whether the Walsh-Healey Act actually has had any significant impact on the nitrogen fertilizer industry. Proponents of OSHA's enactment roundly attacked this Act as insufficient to protect worker health and safety, [97] and people we interviewed in industry and government left us with the impression that the Act may be more than a paper tiger, but not much. [98] The very fact that Congress included in OSHA the requirement that all violations be cited and the provision for minor penalties for first offenses bears witness to the fact that Congress doubted the efficacy of Walsh-Healey.

On the other hand, public standards for ammonia exposure had been set under the Act and were selected largely in consultation with the American Conference of Governmental Industrial Hygienists. Under these circumstances, it is reasonable to believe that the guidelines for manufacturers indicated general agreement on what constitutes reasonable exposure.

Pressure Vessel Safety

Because boiler and other pressure vessel explosions have posed a substantial risk to worker and public safety, boiler standard and inspection regulations have existed at least since the 1920s. Likewise, such private associations as the National Fire Protection Association (NFPA) [99] and the American Society of Mechanical Engineers (ASME) [100] have developed and updated standards for many years. As will be demonstrated below, their standards and those of similar associations are largely the source of public regulation in this area. In this section we will outline the OSHA standards and then those of five states -- California, Iowa, Louisiana, Oklahoma, and

Texas -- to provide an example of how states which
have nitrogen fertilizer manufacturing facilities
have regulated to protect against the risks involved.

OSHA. Consistent with the adoption of national
consensus standards, OSHA adopted a regulation
covering flammable and combustible liquids,
effective February 15, 1972. [101] The source
for the regulation was the NFPA. [102] That
standard, in turn, set standards for pressure
vessel construction in accordance with standards
of the ASME codes. [103]

California. At least since 1929, California
has regulated and inspected boilers and some
pressure storage tanks. [104] The coverage of the
present regulation for boilers is limited to "fired
or unfired pressure vessels used to generate steam
pressure by the application of heat . . . ." [105]
A "tank" is defined to include only "any unfired
pressure vessel . . . used for the storage of air
pressure or liquified gas . . .," [106] except
that a tank "built according to the rules of any
nationally recognized pressure vessel code," is a
tank "for purposes of shop inspection." [107] This
suggests that California regulates many of the
boilers in the ammonia manufacturing process, but
not the reforming furnace or ammonia synthesis
reactor boilers. It also suggests that the
storage tanks used at the manufacturing facility
are subject to inspection by the state.
    Basically, the requirements of the Act are
these:
    (1) No tank or boiler may be operated without
having been issued a permit by the state. [108]
    (2) Installed tanks are to be inspected every
three years, [109] fired boilers internally every
year, and fired boilers externally and unfired
boilers internally and externally at intervals set
by the responsible agency. [110]
    (3) Permits are not to issue unless the boiler
or tank is found to be "in a safe condition for
operation," and the permit must be periodically
renewed. [111]
    (4) Inspections are to be made by "certified
inspectors." An employee of a county, city,
insurer or employer (for purposes of inspecting
his own boilers) is eligible for inspector certification. [112]

(5) Permits may be revoked upon a showing of "good cause," [113] and the agency may order alterations or discontinued use of a boiler or tank it finds to be "in an unsafe or dangerous condition," [114] subject to hearing provisions. [115] Civil restraints and criminal penalties can be obtained against persons violating the Act. [116]

The substance of the Act is not very different from the form originally enacted. The definitions of "boilers" and "tanks" were less specific until they were amended in 1949. [117] Tanks were required to be inspected every two years instead of three until 1955, [118] and all boilers were required to be inspected internally and externally annually until 1961. [119] Other changes seem minor or wholly incidental.

There is no record of litigation under this Act. It will be noted that the statute is extremely vague regarding what must be done to assure that a tank or boiler is "in a safe condition for operation." Our conversation with the appropriate agency led us to believe that the standards applied are essentially those of the relevant ASME and NFPA codes. [120]

Iowa. Iowa first enacted a boiler inspection regulation in 1941. [121] Like California, originally the Act purported to cover all "steam boilers," [122] but it was subsequently amended in 1959 to make clear that it covered only pressure vessels "in which steam is generated by" applying combustion ("fired steam boiler") or vessels in which steam is generated or transferred, but without combustion being directly applied ("unfired steam vessel"). [123] Further, from its inception the Act exempted boilers which are subject to federal inspection and jurisdiction, as well as boilers devoted to agricultural purposes and certain lower pressure boilers. [124] This exemption can make the impact of OSHA on Iowa boiler operators rather vague, since by one fair reading it would suggest that all those subject to OSHA standards and inspections are not subject to this state's regulation. Notwithstanding, the state apparently still does inspect some boilers in nitrogen fertilizer manufacturing firms, but not reactors and other nonfired vessels which are exempted from the Iowa Act. [125]

Under the terms of the Act, a steam boiler and associated tanks and equipment must be inspected, internally and externally, annually by state

52

officials [126] or by "a representative of a reputable insurance company." [127] An exception, permitting only biennial inspections, was made in 1963 for boilers of the size that would be used in nitrogen fertilizer production processes, so long as the boiler is properly supervised and detailed records of its operation are maintained. [128]

Since 1964, the standard for inspecting boilers has conformed "as nearly as possible . . . [to those] formulated by the American Society of Mechanical Engineer's Boiler Code of 1937, as amended." [129] The state boiler inspector informed us that the current ASME code is applied. [130] Failure to comply with the enumerated regulations and rules may result in a fine of not more than $100 [131] or the issuance of an injunction to restrain the use of all alledgedly defective equipment. [132]

Louisiana. Louisiana has provided for regulating and inspecting boilers throughout the state with the exception of the city of New Orleans since 1937. For the most part, the coverage and the enforcement provisions of the Louisiana statute emulate those found elsewhere. Like Iowa, those boilers subject to federal regulations are exempted from the statute. [133] Under the Act, each boiler must be inspected annually, unless, since 1966, it meets standards and recordkeeping requirements that can be met by industrial boiler operators, in which case inspections may be only biennial. [134] Upon inspection, a certificate specifying the maximum pressure that the boiler may be allowed to carry shall issue if the "boiler is found suitable and conforming to the rules and regulations of the commissioner;" but the commissioner reserves authority to suspend an inspection certificate when, "in his opinion, the boiler . . . cannot continue to be operated without menace to the public safety, or when the boiler does not conform to the rules and regulations of the commissioner . . . ." [135] Inspections may be carried out by either representatives of the commissioner of labor or special inspectors acting in a representative capacity. [136]

As originally enacted, rules and regulations pertaining to boiler inspections conformed to the boiler construction code of the ASME whenever possible. [137] In 1975, however, the Act was amended to require full compliance with the ASME

53

Code and "national board standards" whenever the
boiler or pressure vessel required ASME
stamping. [138] The amendment seems to have
created a dichotomy in the application of the Act:
some pressure vessels must conform to the ASME
standards while other need only comply "whenever
possible."

To achieve compliance with the Act, fines
ranging from $25 to $500 or imprisonment may be
assessed for operating a boiler without an inspection certificate or at a pressure exceeding
that specified in the certificate. [139] In
addition to these administrative devices, the
Louisiana Courts have allowed an implied private
right of action to an injured party against an
insurance company, whose inspectors have been
authorized to conduct inspections, for breach of
the duty to competently inspect the boiler which
caused the injury. [140]

Oklahoma. Oklahoma provided for inspection
and testing of steam boilers as early as 1919 [141];
the statutory provisions have remained substantially unchanged since that time. [142] Like
other jurisdictions, Oklahoma exempts certain
classes of boilers and those subject to the jurisdiction of the United States. [143] But in contrast to many states, there is no specific
reference to any national standard for determining
"the positive safety of steam boilers." [144]
Given the State Factory Inspector's authority to
promulgate all necessary rules and regulations to
achieve the positive safety standard, [145]
however, it would not be unlikely for him to look
at the ASME or other national consensus standards.
Support for this position is derived from OLKA.
STAT. tit. 40, §112 which directs the Commissioner
of Labor to prescribe codes for factory inspections
by adopting applicable safety codes of the U.S.
Public Health Service and the American National
Standards Association. [146]

Texas. Texas enacted its first boiler inspection act in 1937; [149] the major provisions
of the Act have remained essentially unchanged.
Consistent with the approach of other states, [150]
Texas exempts boilers under federal control from
state requirements. [151] After 1965, the Texas
legislature extended the scope of its Act by
taking low pressure heating boilers out of the
general exemption category. [152] Two options by

54

the Texas Attorney General have clarified the scope
of the Act's jurisdiction; boilers containing substances other than water are exempt from the
act, [153] but the fact that the steam is
generated by chemicals or hot oils does not remove
the boiler from the requirements of the act. [154]
Although the Texas Act does not refer to any
national standard, [155] and merely conditions a
Certificate of Operation on a finding that "the
boiler is in a safe condition for operation," [156]
the Commissioner of the Bureau of Labor Statistics
is "empowered to promulgate and enforce a code of
rules and regulations in keeping with standard
usage, for the construction, installation, use,
maintenance and operation of steam boilers . . .
and require such devices and safeguards and other
reasonable means and methods to insure safe operation of steam boilers . . . ." [157] Any such rule
or regulation must first satisfy procedural safeguards. [148] In addition to issuing a certificate
of operation, the commissioner requires that all
boilers subject to the act be registered. [159]
Inspections are conducted on an annual basis unless
the boiler owner adopts continuous water treatment
to resist corrosion and maintains a set of
records, [160] in which case the interval between
inspections may extend for as long as twenty-four
months. [161] If a boiler is inspected and subsequently insured by an insurance company, no state
inspection is required. [162] To ensure compliance with the provisions of this Act, the
commissioner may obtain a temporary restraining
order against continued use of the boiler, [163]
require repairs, and file a misdemeanor action,
punishable by fine up to $200 or imprisonment not
to exceed sixty days. [164]

Conclusions. Boiler regulation has existed for
a long time but much of the emphasis of the regulation does not seem to be directed against the
large manufacturer with major industrial boilers.
Accordingly one can surmise that the state did not
consider it necessary to intensively regulate and
inspect such boilers to ensure proper maintenance
and operation.
Standards that are applied by the states
generally are not an additional burden for the
boiler operator since they tend to track the ASME
codes. Indeed, at least one author has suggested
that the standards are not high enough because the
engineer-designers tend to address themselves to

whether they have complied with the codes rather
than to whether they have designed a safe
boiler. [165] It is, of course, possible that the
ASME standards are unreasonable or that they
unreasonably restrict innovation. If so, the state
regulations may contribute to that restriction
since they give standards the force of law. This
cannot be more than a minor effect, however, for
two reasons. First, as written, the state
standards either automatically take into account
any amendments of the ASME codes or are capable of
taking them into account upon administrative
action not requiring an act of the state legislature. Second, the ASME codes are obeyed for
reasons other than the requirements of the boiler
regulation. A failure to comply with the code can
raise questions about a design engineer's professionalism and may cause a purchaser to doubt
the craftsmanship of the noncomplying product.

The required inspections are the other feature
of these regulations that might be unreasonable
or unreasonably restrict innovation. If a state
were to take the view that the internal condition
of a manufacturer's boilers should be inspected
on a surprise basis, the manufacturer could lose
valuable production time. There is no readily
apparent reason for requiring surprise inspections,
and by all indications inspections are done on a
scheduled basis.

There may be another problem if internal
inspections of boilers are required more frequently than the times that the boilers are brought
down for scheduled maintenance. Under existing
laws this is probably not required although in
some states it may once have been. That may
explain why California and Iowa decreased the
frequency of some required inspections.

One interesting question remains unresolved.
Several states exempt from their own inspection
and certification rules those boilers that are
subject to federal jurisdiction. When enacted,
these exemptions were made with the federal inspection of railroad boilers in mind. However,
OSHA now has regulations covering boilers in
chemical plants. As noted above, those regulations
do not necessarily include the intensive inspection
program characteristic of the state laws. The
question is whether the lesser federal regulations
will not supersede the state regulation, thereby
relaxing the regulation of pressure vessels.

Workmen's Compensation

Prior to the development of workmen's compensation, a disabled worker was required to bear the risk of his injuries unless he could prove his employer's negligence and rebut his employer's proof (1) that he himself was negligent, (2) that he had knowingly assumed the risk of injury by having taken the job, and (3) that the injury was caused by a fellow employee. By the turn of the century, this pattern began to change with the enactment of the early workmen's compensation laws, designed to provide compensation for work related injuries regardless of the comparative negligence or risk assumption of employees. Not until 1917 did the Supreme Court of the United States rule that compelling employers to participate in such programs was constitutional, [166] and by that time a series of statutes had been enacted which allowed workmen's compensation program participation at the election of the worker and employer, but did not mandate it. For example, such programs were enacted in California in 1911, [167] in Iowa in 1913, [168] in Louisiana in 1914, [169] in Oklahoma in 1915 [170] and in Texas in 1917. [171] By now, all states either require participation in workmen's compensation programs or allow election but impose serious legal disabilities on those who choose not to participate. The programs enacted in the 1910s were criticized because the compensation allowed was inadequate, an insufficient number of types of incidents were covered, and too few workers were within the ambit of the rules. From the point of view of many, that situation has not yet been satisfactorily resolved.

The history of workmen's compensation is well documented and need not be repeated here. [172] Suffice it to say that, among the major issues surrounding these programs, several are particularly important for nitrogen fertilizer manufacturing firms. Originally, workmen's compensation programs in most states were exclusively intended to compensate workers for injuries caused by accidents. Occupational diseases were not covered until these programs were extended or supplemented by enactment of occupational disease statutes. This occurred, for example, in Iowa in 1947, [173] in Louisiana in 1952, [174] in Oklahoma in 1953, [175] and in Texas in 1947. [176] Under the circumstances, it became significant whether the injury resulting from exposure to ammonia, nitrous oxide, or nitric acid was defined as an accident or an occupational disease. An accident has traditionally been defined as a sudden, unexpected injury of particular time and duration. [177] If one were to interpret such a definition liberally, one would anticipate that the injury resulting from ammonia or nitric acid exposure would be characterized as an accident. And, at least for some years prior to enactment of the occupational disease statutes, many courts were predisposed to extending the coverage of workmen's compensation laws by liberally interpreting "accident." [178] Injury caused by mechanical accidents, such as failure of reciprocating compressors or vessels, would, of course, be considered accidents without debate.

Another workmen's compensation issue having a clear impact on the nitrogen fertilizer industry involves the amount of the compensation awarded for injuries. The generally accepted view is that benefits are insufficient in some categories of compensation. [179] Finally, whether the workmen's compensation laws serve to encourage employers to use safe practices is a significant question. Presently, about 65 percent of the nation's employers under workmen's compensation are provided benefits by private insurers, while 23 percent are covered by state-operated funds and 14 percent are self insured. [180] Many in the private and state funds are charged premiums that have no relationship to accident experience. Under this circumstance, the economic incentive to provide safe working conditions is reduced.

Workmen's compensation has been a fact of manufacturing life for quite a long time. Injuries which are likely to occur in nitrogen fertilizer manufacture often are covered by workmen's compensation even when such programs do not cover occupational diseases. Those injuries, if any, that were not covered originally have become covered by enactment of provisions covering occupational diseases. Finally, it is not clear that workmen's compensation laws increased the cost of doing business for manufacturers, since the introduction of these laws eliminated worker's rights to recover for accidental injury under common law negligence.

3.4   ENVIRONMENTAL PROTECTION REGULATION

Water Pollution Control

When Congress enacted the Federal Water
Pollution Control Act Amendments (FWPCA) in
1972 [181] it drastically restructured the nation's
approach to restoring and maintaining the chemical,
physical, and biological integrity of the nation's
waters. To achieve by 1983, wherever attainable,
a water quality which would provide for the protection of fish, shellfish, and wildlife, and to
develop technology necessary to eliminate discharges of pollutants, Congress adopted a fresh
approach. [182] Whereas prior statutory schemes
had relied upon water quality standards as the
primary means of pollution control, the FWPCA
supplemented these standards with the application
of control technology to the sources of
pollution. [183] By adopting effluent limitations
which would prohibit discharges of pollutants in
greater concentration or volume than permitted by
the limitation and which might be adjusted to
achieve the desired level of water quality,
Congress sought to realize the goal of eliminating
the discharge of pollutants into navigable waters
by 1985. [184] To achieve the goal, Congress
ordered the Environmental Protection Agency (EPA)
to publish a list of categories of sources for
which effluent limitation standards would be
developed. [185] Fertilizer manufacturing was
among the twenty-seven categories specifically
enumerated by Congress. [186]

     Control Before the FWPCA. Federal involvement
in water pollution control began long before
ecology was a household word. Although measures
that indirectly affected the quality of the
nation's waters were enacted as early as the first
Congress, [187] the first modern, major legislative effort was the Water Pollution Act of
1948. [188] While the Act operated on the premise
that the primary responsibility for curbing
pollution lay with the states and provided for
financial assistance to states and municipa-
lities, [189] it also authorized federal abatement
actions against polluters of interstate waters;
that is, waters that flow across or form part of
state boundaries. [190] Because federal

enforcement required lengthy preliminary hearings and the consent of the state in which the violation occurred, the Act failed as a mechanism for curbing pollution. [191]

Between 1948 and 1966 the Federal Water Pollution Control Act was amended four times -- 1956, 1961, 1965, and 1966. The Water Pollution Control Act Amendments of 1956 [192] eliminated some of the procedural delays and the states' veto power over federal enforcement actions, but inserted additional delays and narrowed the jurisdiction of the federal government's enforcement powers. [193] The 1961 amendments [194] not only restored the jurisdictional formulation developed in the 1948 Act, but extended it to include intrastate waters that were not tributaries of interstate waters. [195] Additionally, the federal government could enter intrastate cases at the request of the governor. [196] Finally, federal abatement authority was expanded to cover intrastate pollution when the health or welfare of persons within the affected state was endangered. [197] Once again, however, the exercise of federal authority was contingent upon the state's invitation.

When Congress passed the Water Quality Act of 1965, [198] the federal approach to water pollution control took a new twist. For the first time, the Act established water quality standards. [199] Although the initiative rested with the states, if they failed to adopt water quality standards the Secretary of HEW could promulgate his own standards. [200] This legislation did not produce any measurable improvement in many areas of the country for several reasons. First, water quality standards are not designed for use primarily as an enforcement device, but instead are used to promote the orderly development of water resources. [201] Therefore, their impact on water quality was slow to develop. Second, the Act provided little guidance to the states engaged in developing water quality standards. [202] Enforcement was also hampered by the requirement that the state give consent before the federal government instituted abatement action. [203] The Clean Water Restoration Act of 1966 [204] did little to improve the enforcement difficulties, but it did authorize the government to require an alleged polluter to provide information concerning the nature of his pollution problem. [205]

The primary problem with all of these pieces of legislation was their lack of an effective means for ensuring enforcement. Their failure might be measured against the record: from 1948 to 1967 only one abatement action was brought to trial. [206] In short, it appears that the federal water pollution control legislation prior to 1971 had little regulatory effect on fertilizer manufacturing.

The Water Quality Improvement Act of 1970 [207] introduced a new approach for achieving compliance with established water quality standards. Section 21 of the Act required any facility which introduced discharges into the navigable water of the United States to secure a federal permit. [208] Before a permit would be issued, the facility must demonstrate that its discharges would not violate applicable water quality standards. Given the brief interval between the 1970 Act and the 1972 Amendments which set effluent limitations from point sources, however, it is difficult to identify an impact from the 1970 Act independent of the 1972 Amendments.

An Overview of FWPCA. The development of effluent limitations involved three phases. [209] The first concerned standards for facilities that are new sources, that is, facilities going into operation after the passage of the Amendments. [210] The second required that by 1977 existing point sources meet the effluent limitation achievable after the application of the best practical control technology currently available. [211] The third required adherence to effluent limitations attainable by utilizing the best available technology economically feasible by 1983. [212] When technologically and economically possible, the FWPCA empowered EPA to require elimination of all pollution discharges. [213]

To ensure compliance with the effluent limitations, the Amendments declared that "the discharge of any pollutant by any person shall be unlawful" unless otherwise permitted. [214] One of the statutory exceptions involves discharges allowed under a permit issued by EPA or its state representative. [215] Unless the applicant can show his particular situation is sufficiently unusual to justify special treatment, to secure a permit one must meet the effluent limitation standards which are gradually tightened for 1977, 1983, and 1985. [216] Thus, the permit program

61

provides an enforcement mechanism to control the discharge of pollutants and to guide the transition to the 1977, 1983, and 1985 effluent limitation goals. [217]

Ambiguity in the statutory language [218] spawned considerable litigation over whether the EPA-promulgated effluent limitations were merely guidelines to assist the individuals responsible for considering permit applications. That litigation lasted several years, resulted in varying decisions among the Federal Circuit Courts of Appeals, [219] and was ultimately resolved by the Supreme Court in 1977. [220] This situation produced uncertainty about the operation of the FPWCA for nitrogen fertilizer manufacturers as well as all others. In the course of the litigation, industry argued that EPA could not promulgate general standards which could be enforced as a condition of each permit issuance. Rather, according to this view, EPA was required to make a case-by-case review of the limitations to be imposed upon each permit applicant, relying on the guidelines only as a basis for standards for each specific case. [221] Such a result would assist industry because it would ensure that guidelines stated in general terms would not be automatically applied without attention to the peculiarities of each case. It would also have the effect of retarding the rate at which EPA could implement the Act, perhaps cause EPA to clear permit applications too hastily, and force EPA to devote staff resources to permit-processing instead of other activities.

Apparently to avoid these deleterious effects on its performance and to establish its authority, EPA took the view that it had the power to adopt guidelines as the standard it would require of all permit applicants not obtaining a variance. [222] Since variances would be allowed only for very special cases, the effect of the EPA's position was to apply the guidelines broadly across industries.

Five Federal Circuit Courts of Appeals ruled on the matter before the Supreme Court settled the question. Four (the Second, Third, Fourth, and Seventh Circuits), and then the Supreme Court, [224] ruled in favor of EPA. Industry-wide standards may now be imposed for all standards -- those for new point sources and those to be in operation in 1977 and 1983, respectively.

Fertilizer Manufacture Regulation Under the
FWPCA. On April 8, 1974, EPA announced promulgation of effluent limitations for fertilizer
manufacturing. [225] Separate limitations were
deemed appropriate for different segments of the
industry. The classification of the industrial
segments -- ammonia, ammonium nitrate, nitric
acid and urea -- largely rested on the individual
water treatment technologies associated with each
component. [226] For instance, in an ammonia
manufacturing facility, adequate treatment of
the primary waste water constituent, ammonia,
requires treatment separate from other fertilizer
manufacturing operations which might be taking
place; [227] treatment of the waste water from a
urea facility by urea hydrolysis necessitates
isolating the urea waste stream from the waste
streams associated with other products; [228] and
the flash vaporization of water from nitric acid
to make ammonium nitrate poses special treatment
problems that must be considered separately. [229]
    Industries, like fertilizer manufacturing,
that produce products in several of the classifications in the same plant have objected to such
classifications on the grounds that they are
artificial distinctions unrelated to the actual
production activity of the plants. Several courts
of appeals have sustained EPA's classifications,
holding that requiring standards to reflect each
individual production configuration would be
impractical and contrary to legislative intent.
[230]
    The criteria considered in establishing
effluent limitations guidelines are outlined in
the development document which preceded the final
promulgation of the regulations. [231] By 1977,
manufacturers were to have met effluent limitations
levels attainable through the application of the
best practical control technology. [232] The
Administrator determined that level of technology
by surveying the performance of existing plants,
giving special attention to the performances of
plants deemed to be exemplary, though varying in
size, age, and process employed. If no exemplary
plants were surveyed or existed, EPA relied upon
the level of technology which represented the
state-of-the-art unit operations commonly employed
in the chemical industry. [233]
    Under the regulatory scheme that requires
application of the best practical control technology currently available, emphasis was placed

63

on end-of-pipe treatment, but compliance may be
by in-process treatment. [234] Other factors considered in establishing the effluent limitations
include: the cost/benefit ratio, which is computed from the costs of applying the technology
and the benefits from the expected level of
effluent reduction; the size and age of the
facilities involved; the processes employed; the
engineering aspects of applying various types of
control technology; and the consequential nonwater
environmental impact of implementing the limitations. [235]

For each manufacturing segment of the industry,
EPA first identified the primary pollutant discharged in the manufacturing process. All of the
industrial segments discharge wastes containing
low and high pH and nitrogen, though the form of
the nitrogen depends upon the individual manufacturing process. [236] In the case of ammonia
manufacturing, the nitrogen takes the form of
ammonia. [237] In addition to the pH and nitrogen
wastes, nitrogen fertilizer manufacturing wastes
may contain a number of other pollutants, including
dissolved and suspended solids, oil and grease,
chromium, zinc, iron, and nickel. [238] EPA has
not established effluent limitations for those
substances, however, because it believes that
treatment of the primary pollutants would effect
sufficient removal of the secondary substances.
Also, a lack of supporting data militated against
establishing such effluent limitations.

The standards for the ammonia manufacturing
subcategory are unique insofar as they have remained
substantially unchanged. As originally promulgated,
the amount of ammonia discharged into waste water
could not exceed 0.0625 kg/kkg (lb/1000 lb) of
product for the maximum average of daily average
levels for a period of thirty consecutive days.
[240] That level of discharge represents a
slight tightening over the level proposed in the
development document, 0.063 kg/kkg; no explanation
was given for the variation. [241] In addition
to the thirty day average value, EPA issued a
standard allowing a maximum one day discharge level
equal to twice the thirty day value, or 0.125
kg/kkg. [242] These figures are based upon the
results experienced by plants presently operating
and using control technologies based on ammonia
stripping by air and/or steam. [243]

On June 23, 1975, the Administrator amended
these standards upward to allow a maximum one day

discharge level based on three times, rather than
twice, the daily values for a thirty day period.
[244] As EPA explained the change, "this regulation is being made as a result of receipt of
new information from industry representatives
which indicated that the variability in treatment
performance was greater than initially indicated." [245]

Aside from this adjustment, the only other
major change in the scope of the ammonia regulations involved the elimination of an effluent
limit for oil and grease. The proposed regulations
for ammonia manufacture had set a maximum one day
level of oil and grease discharge of 0.025 kg/kkg
of product, and a 0.0125 kg/kkg of product daily
average over a thirty day period. [246] This
standard was excluded from the regulation as
finally promulgated because the "limits were within
the range of questionable reproducibility for the
standard method of analysis." [247] Although the
EPA recognized the option of increasing the limit
to a point where analysis would be more reliable,
it felt that the better approach would be to base
limitations for this pollutant on water quality
criteria. [248]

To meet the effluent limitation levels set out
in the regulations, the EPA listed the various
technologies which complied with the levels.
Control of oil and grease could be achieved through
strategic placement of drip pans and application
of gravity API separators. [249] The principal
mode of treating the nitrogen waste water is
ammonia stripping by air and/or steam. Other control technologies adaptable for this particular
pollutant include biological nitrification/
denitrification and selective ion exchange for
ammonia subsequent to the stripping. [250]
Improved stripping technology, currently under
development, served as the foundation for the
effluent limitations employing the best available
technology economically feasible in 1983. [251]
(See Chapter 5 for further discussion of these
and other technologies for the control of
effluents.)

In the comment period that followed the initial
effluent limitation proposal, representatives of
the ammonia manufacturing industry made several
suggestions and criticisms of the regulations.
The effectiveness of their input might be judged
by EPA's willingness to abandon the standards for
oil and grease in the wake of criticism. [252]

On several other points, however, the agency re-
fused to amend its standards. In response to the
observation that the standards for new sources were
not as stringent as the standards relying on the
best available technology economically achievable,
EPA noted that the new source standard applied to
any facility constructed after the passage of the
1972 FWPCA whereas the best available technology
economically achievable standard would not be
implemented until 1983, and therefore, could
encompass technology that was yet to be perfected.
As promulgated, the performance standards for new
sources are more relaxed than the 1983 standards,
but more stringent than the 1977 standards. [253]
    A criticism voiced by several industrial
spokesmen concerned EPA's failure to make allowances
for leaks and spills occuring during the course of
manufacturing. EPA contended that good housekeeping and efficient operations would resolve and
minimize this problem. [254] In addition to this
rationale, the development document suggested that
ordinary leakage and spillage problems could be
handled under the one day maximum discharge level
permitted under the regulations. [255]
    Another criticism challenged EPA's reliance
on the air stripping control technology as the
principal means for achieving the 1977 limitations.
That technology was labelled inappropriate because
the ammonia dispersed to the air by this process
eventually reappears in the water cycle. In contrast to air stripping, steam stripping or urea
hydrolysis do not produce any air pollution
problem. The agency dismissed these objections
noting that there are currently no EPA air quality
standards for ammonia and that the concentration
level of ammonia resulting from air stripping was
below the threshold level of human odor per-
ception. [256]

    Litigation of the Fertilizer Manufacture Ef-
fluent Limitation Standards. While the industry's
compliance with the relevant effluent limitation
guidelines for nitrogen fertilizer manufacturing
has been generally high, [257] the regulations have
not escaped legal challenges. In Vistron Corp. v.
Train, [258] twelve firms representing the major
manufacturers of nitrogen fertilizer filed a
petition for judicial review of the regulations
with the United States Court of Appeals for the
Sixth Circuit. The litigation focused on EPA's
promulgated effluent limitation guidelines for the

nitrogen fertilizer industry and in particular,
those setting zero discharge standards. [259]
Industry argued that absolute prohibitions of discharges were unreasonable because discharges
resulting from accidents occurred from time to
time. [260] Before the court acted upon petition
for judicial review, the parties, with the court's
approval, settled the dispute. The settlement
required the industry representatives to supply
data concerning the need for a special discharge
amendment to the nitric acid regulation and to
give EPA information regarding the need for
attaching a leaks and spills amendment to all
nitrogen fertilizer regulations. The settlement
also required that EPA consider the data submitted
and advise the petitioners whether amendments would
be promulgated. [261]

Of the four manufacturing categories, only
the nitric acid regulations set a zero level of
discharge. [262] In accordance with the terms
of the Vistron agreement, EPA amended the regulation to allow some nitrogen discharges in nitric
acid production. [263] If industry practiced good
housekeeping and observed efficient operations
and maintenance, EPA believes that any accidental
discharge could be kept to a low maximum standard.
Moreover, the agency viewed this approach as
obviating the need for any special regulation for
leaks and spills. [264] As for the other
subcategories -- ammonia, ammonia nitrate, and
urea -- EPA concluded that leaks and spills
admendments were not necessary. [265]

EPA opposition to proposals calling for specific
leaks and spills amendments is longstanding. In
the Development Document which preceded the promulgation of the regulations, it discussed the
sources of leaks and spills and the means of prevention. [266] Defective pump seals, bad valves,
and spillage occurring in the course of loading
or transporting fertilizer, especially fertilizer
in solution form, account for most of the problems. [267] While recognizing these facts,
however, it did not allude to the problem in the
context of its examination of various control
technologies, but merely announced that best practical control technology currently available
includes leak and spill control as well as good
housekeeping. EPA referred to the Manufacturing
Chemists Association's publication, "Guidelines
for Chemical Plants in Prevention, Control and Reporting of Spills" for containment techniques. [268]

During the comment period following the proposal of effluent limitation guidelines, several
commentators took the agency to task for not taking
leaks and spills into account when establishing
the proposed guidelines. [269] Responding to this
criticism, EPA disagreed: good housekeeping
practices, efficient operation and prompt maintenance would minimize waste from leaks and spills.
Moreover, it argued, the resulting waste water
could be segregated from otherwise contaminated
streams and recovered for dry disposal or reused
in the production process.

It is not clear whether the statute, as
interpreted by the courts, permits EPA to rely
upon good housekeeping and maintenance as the
technology to be used for complying with its
standards. In a challenge to effluent limitation
guidelines developed for the plastics industry,
the Fourth Circuit Court of Appeals rejected
regulations because the Court found it "impossible
to determine the reasonableness of the 1983
limitations when a major element in the calculation . . . is not made known by the Agency." [270]
If one argues that EPA has failed to set out the
calculations that support reliance on good
housekeeping as the prime means for controlling
leaks and spills, it is quite possible that a
reviewing court would remand the regulations for
the reasons given by the Fourth Circuit.

The legal wrangling aside, an ultimate question
is whether EPA's effluent limitations are feasible.
That is a matter of continuing debate. To
establish its standards, in 1973, EPA had fifteen
plants surveyed. [271] It concluded that the
limitation guidelines were met by some exemplary
plants and could be met by any nitrogen plant
employing best practical control technology
currently available. It did not, however, identify
which plants were currently meeting the limitations. [272] Throughout its supporting documentation, EPA referred to one control technology
or another as being successful -- without
specifying the levels of effluent discharge
achieved through that technology. We examined the
public documents file of the raw data compiled in
the survey. We found indications that a few
plants were meeting pH levels but no evidence that
any plants were meeting the nitrogen discharge
standards at that time. [273] However, the absence
of that evidence is not conclusive, since the data
on file was sketchy and occasionally illegible.

Air Pollution Control

Although federal involvement in air pollution
control began in 1955 with the passage of the Air
Pollution Control Act, [274] the quality of the
nation's air continued to deteriorate. Subsequent
legislation in the 1960s [275] failed to produce
any significant improvement for several reasons:
lack of vigorous enforcement, the absence of any
compliance deadlines, and the federal government's
failure to publish guidelines to aid the states in
their efforts to establish implementation plans.
[276] To remedy these deficiencies and to
institute, a cohesive approach for curbing air
pollution, Congress enacted the Clean Air Act
Amendments of 1970.

This legislation was designed to increase
federal involvement in air pollution control. It
required EPA to establish national standards of
performance for new stationary sources; [277]
required EPA to identify and set national standards
for hazardous air pollutants; [278] and allowed
EPA to independently issue compliance orders or
file suit. [279] If EPA believed that a state
had failed to enforce its plan properly, it could
intervene and enforce it. [280] Moreover, the
Amendments pressured the states to step up their
enforcement efforts through the establishment of
national air quality standards. [281] Each state
was required to develop an implementation plan
which took the federally-promulgated standards
into account, [282] though the states were free
to establish stricter standards. [283] State
compliance with the federal standards was further
assured by the imposition of attainment deadlines [284] and a provision for federal review of
mandatory state implementation plans. [285] If
the state failed to develop a satisfactory plan,
EPA was empowered to prescribe and enforce an
implementation plan. [286]

Two Supreme Court decisions have clarified the
scope of EPA's review powers. In Train v. NRDC,
[287] the Supreme Court held that states could
grant variances to polluters if attainment and
maintenance of the national air quality standards
are not threatened. And in Union Electric Co. v.
EPA, [288] the Supreme Court held that EPA was not
permitted to consider economic and technological
feasibility in reviewing a state's implementation
plan. While the Union Electric decision permits
EPA to embark upon a technology-forcing regulatory

scheme, the variance loophole created by Train v. NRDC may blunt the impact of federal regulations.

In 1971 and 1973, acting under its mandate to establish national air quality standards, EPA promulgated national primary and secondary ambient air quality standards for a number of pollutants, including particulate matter, hydrocarbons, and nitrogen dioxide. [289] Ammonia manufacture involves potential air emissions of these pollutants (including emissions resulting from the removal of nitrogen from water effluents by the steam stripping process). However, as far as we can discern, air pollution regulation has not had a significant direct impact on ammonia manufacture for several reasons. First, air quality standards are directed toward ambient air quality in geographic regions and are not applied to specific point sources. [290] Thus, those ammonia manufacturing facilities which are located in rural and other areas where air quality is relatively good are not matters of high priority concern for EPA. Second, operation of air pollution control regulations for existing air pollution sources has been left to the states, which are authorized to grant variances for individual sources from complying with broad standards. [291] Third, most modern ammonia manufacturing facilities are located in a complex of facilities that also manufacture ammonium nitrate, nitric acid, and urea. The air pollution of those operations far overshadows pollution resulting from ammonia manufacture. [292]

Another major component of the Clean Air Act Amendments sets air emission limits on new sources of pollutants for a number of industries. [293] These stationary source restrictions are not applied to ammonia manufacture, although they are to nitric nitric acid manufacture. [294] Of course, since new nitric acid plants are usually associated with ammonia production, nitric acid regulation presents an incentive to limit air emissions from the entire complex. This might suggest the reason for the pervasive reduction of air pollution at new nitrogen fertilizer manufacturing facilities. [295]

State regulatory activity in the air pollution sphere parallels the federal scheme in many respects. Consequently, it is not surprising to find that no state regulations are specifically directed against ammonia plants. Several states, including Minnesota and Texas, have devised special regulations for

70

nitric acid plants. [296]   Georgia regulates discharges of particulate matter from fertilizer
plants, but the regulation appears to be aimed at
blending and prilling processes -- both of which
are foreign to ammonia manufacturing. [297]
      Thus, we conclude that federal and state air
pollution control activity through 1976 had only
a minimal impact on ammonia manufacturing. This
is not to say that matters will necessarily remain
unchanged. Over the past seven years, EPA has
added several pollutants to those for which
national ambient air quality standards have been
devised. [298]   If EPA were to add ammonia to the
list, the entire nitrogen fertilizer industry,
including ammonia manufacturing, would find itself
heavily regulated on air pollution matters. Also,
the future implications of new rules under the
Clean Air Act Amendments of 1977 [299] may change
the nature of ammonia manufacture even more. [300]

4. Ammonia Process Technology

4.1 INTRODUCTION

In this chapter we describe the technologies used to produce synthetic ammonia since 1913, the year of the first operation of the Haber-Bosch synthesis from hydrogen and nitrogen. For convenience, we consider ammonia plants in terms of two major components -- a means of producing hydrogen and nitrogen, and a means for reacting them to produce ammonia. The major technologies were distinguished from one another in the early period of the industry according to the means of effecting the ammonia synthesis. As the industry matured, innovation moved to the arena of hydrogen production. In more recent years major innovations were made by discarding this paradigm of two separate elements and developing technologies that optimize over the entire plant.

In Section 4.2 we discuss the technology of ammonia synthesis, based on the Haber-Bosch process and its derivatives. In Section 4.3 we discuss the technology of hydrogen production in a general way. Finally, in Section 4.4 we discuss in some detail each of the major processes which have been used commercially. The commercial processes are organized and discussed according to the means of hydrogen production. Steam reforming of natural gas is divided into two technologies according to whether reciprocating or centrifugal compressors are employed.*

---

*It should be noted that any taxonomy of ammonia process technology is somewhat arbitrary. We follow the practice of the literature which focuses on ammonia synthesis differences in the early

4.2 AMMONIA SYNTHESIS

Today virtually all ammonia is produced commercially by direct synthesis from hydrogen and nitrogen by means of the Haber-Bosch process. The very simple chemistry of the process can be described by the chemical equation of equilibrium:

$$N_2 + 3H_2 \rightleftarrows 2NH_3$$

For reasons discussed below, the reaction is carried out in the presence of a catalyst at high temperature and pressure using a gas mixture which contains relatively small amounts of certain impurities.

The optimum operating temperature, pressure, and catalyst for ammonia synthesis are dependent upon two major concerns: the equilibrium conversion of N_2 and H_2 to NH_3, and the rate at which that conversion is reached. The chemical relationship between the three gases is an equilibrium one, which means that after a sufficient time has elapsed, each of the gases will be present at a fractional concentration that depends only on pressure and temperature. Haber and others made careful measurements of this equilibrium and found that the concentration of NH_3 is enhanced by low temperature and high pressure, as shown in Table 4.1 taken from Slack and James. [1]

That this should be the case follows from two facts: first, assuming ideal gas behavior, the reaction of one volume of N_2 and three volumes of H_2 produces only two volumes of NH_3. Reactions with volume decrease are enhanced by high pressure.

---

period, on hydrogen production in the middle period, and on plant integration and scale in the later period. One could also organize as to feedstock (which is intimately bound to hydrogen production technology) or as to the pressure levels used in the hydrogen production section or in the ammonia synthesis chamber. Finally, one can organize the technologies according to historical periods: coke-water gas and by-product gas (1921-1940), low pressure reforming of natural gas (1941-1953), high pressure reforming (1953-1963), and centrifugal compressors (1966-present).

TABLE 4.1
Equilibrium Concentrations of Ammonia in
Ammonia/Nitrogen/Hydrogen Gas Mixtures

Temperature °C	Ammonia Concentration (% by volume) Pressure (atm)			
	1	30	100	200
200	15.3	67.6	80.6	85.8
300	2.18	31.8	52.1	62.8
400	0.44	10.7	25.1	36.3
500	0.129	3.62	10.4	17.6
600	0.049	1.43	4.47	8.25
700	0.0223	0.66	2.14	4.11
800	0.0117	0.35	1.15	2.24
900	0.0069	0.21	0.68	1.34
1000	0.0044	0.13	0.44	0.87

Source: Adapted from Slack and James [2], p. 37, by
        courtesy of Marcel Dekker, Inc.

Second, the reaction of N_2 and H_2 is accompanied by
release of energy as heat, and such reactions are
favored by low temperatures.
    The rate of reaction between N_2 and H_2, how-
ever, is quite slow at low temperatures. Thus,
while at low temperature good equilibrium conversion to NH_3 is achieved, it is achieved only
after a very long time. This problem can partly be
overcome by the use of a catalyst -- a substance
which increases the rate of a reaction while having
no effect on the final equilibrium concentration.
Haber, Bosch, and others examined at least 2500
substances before developing a successful catalyst
in their early work at BASF in Germany.* [3]
    A modern ammonia synthesis unit is a con-
tinuous process, operating at 100-200 atm and

---

*Both men received Nobel prizes for their work on
ammonia synthesis: Haber in 1918 and Bosch in
1931.

300-500°C with a catalyst consisting of iron oxide
containing small amounts of potassium and aluminum.
It is shown schematically in Figure 4.1, taken from
Slack and James. [4]

Much of the early innovation in ammonia production centered around design of the synthesis
reactor and selection of temperature, pressure, and
catalyst. A number of processes were developed
following the Haber-Bosch process, including the
Mont-Cenis, Claude, Fauser and NEC. [5] They
differed primarily in the details of internal
reactor construction, catalyst nature and shape,
and operating pressure level. It has been suggested that many of these developments were made
to circumvent the patent position of BASF. [6] A
summary of the features of the major synthesis
processes as of 1964 from Sauchelli [7] is shown
in Table 4.2.

It is apparent from Table 4.1 that a relatively
small fraction of the input N_2 and H_2 appear as
product NH_3 in the neighborhood of ordinary
operating conditions, 500°C and 150 atm. Since
considerable effort is necessary to produce high
purity H_2 and N_2 for the reaction, it is worthwhile
to separate the product NH_3 and recycle the
remaining H_2 and N_2 back through the reactor. This
practice has been followed in all but the original
Claude process, which operated at very high
pressures (about 1000 atm).with several reactors
in a series-parallel arrangement to produce a
product gas containing about 80 percent NH_3. [8]
Unreacted gases are burned as fuel in the Claude
process rather than recycled as feedstock.

Ammonia is recovered from the product stream
by cooling it with water or by refrigeration,
depending upon the pressure level. At high
pressures, ammonia can be liquified and removed at
temperatures achievable by ordinary cooling water.
At lower pressures, refrigeration of the product
is required. Thus, an economic trade-off exists
between low pressure operation, which requires
expensive refrigeration, and high pressure
operation, which requires expensive compression.

The feed stream to the synthesis reaction
invariably contains not only H_2 and N_2, but also
smaller amounts of methane, water vapor, argon and
other impurities. The water vapor condenses out
of the product with the liquid ammonia product
stream, but the gases do not. To avoid build-up
of the concentration of these gases, a "purge"
stream must be diverted from the main recycle gas

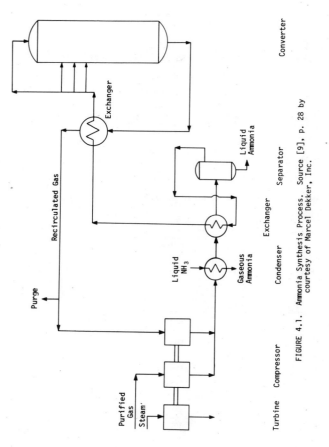

FIGURE 4.1. Ammonia Synthesis Process. Source [9], p. 28 by
courtesy of Marcel Dekker, Inc.

77

TABLE 4.2
Comparison of Some Ammonia-Synthesis Processes

System	Pres-sure Cat-egory	Basic Converter Features	Approximate Operating Conditions	
			Pressure (psi)	Peak Temp Range (°F)
Haber-Bosch (original)	low	Single catalyst charge with feed preheat.	3000	930-1100
Claude (original)	high	Single catalyst charge with preheat of feed in converter annulus. Series-parallel operated converters.	15,000	930-2100
Claude (present) Grande Paroisse	high	Single catalyst charge with heat exchange in catalyst bed.	5000-9500	1000-1100
Casale	high	Single catalyst charge with preheat via internal heat exchange. Use of special catalyst.	9000	850-1000
Mont Cenis	low	Internal heat exchange in catalyst bed. Use of high activity low temperature catalyst.	1500-2400	750-800
TVA	low	Single catalyst charge with countercurrent cooling tubes. Interchanger to preheat feed.	3700-5200	900-1000
Chemico	low	Single catalyst charge with co-current cooling tubes. Interchanger to preheat feed.	3700-5200	900-950
Fauser-Montecatini	low	Interchanger to preheat feed. Catalyst temperatures controlled by steam generation between catalyst beds.	4000-5000	950
Kellog	low	Interchange to preheat feed. Catalyst temperatures controlled by introduction of cold feed between catalyst beds.	3500-4700	900-950
UHDE	low and high	Interchanger to preheat feed. Catalyst temperatures controlled by introduction of cold feed between catalyst beds.	4600-6400	950
Lummus	low	Interchanger to preheat feed. Single catalyst charge with counter-current cooling tubes.	4000-5000	930-950

*Conversions estimated from trade information.
Source: Sauchelli [10], pp. 78-79, by courtesy of American Chemical Society.

Approximate Hydrogen Conversion* (%)	Approximate Converter Ammonia Concentrations* (mole %)		Method of recovering product	Method of Recycling
	Input	Output		
9-18	nil	5-10	Water scrubbing.	compressor
80 (overall)	—	25	Water cooling and single stage condensation followed by water scrubbing.	none
30-34	3-4.5	20	Water cooling and single stage condensation. Also water cooling and refrigeration for two stage condensation.	compressor
30	3-5	23	Water cooling and single stage condensation.	ejector
9-20	—	5-12	Water scrubbing and also temperature refrigeration.	compressor
25	5	17.5	Water cooling and refrigeration. Two stage condensation.	compressor
25	3	17	Water cooling and refrigeration. Two stage condensation.	compressor
30	1.5	20	Water cooling and refrigeration. Two stage condensation.	injector
25	3	17	Water cooling and refrigeration. Single and two stage condensation	compressor
25-30	3	17-20	Water cooling at high pressure. Water cooling and refrigeration at low pressure.	compressor
25	3	14-18	Water cooling and refrigeration. Two stage condensation.	compressor

stream to remove a portion of the inerts on a continuous basis. This purge stream contains not only the impurities but a significant amount of the N_2, H_2, and NH_3 as well. In a few plants, it has been the practice to recover the purge stream ammonia as well as the hydrogen, nitrogen, and argon for sale or use elsewhere. More frequently, however, the purge stream is used as a fuel gas for the hydrocarbon reforming furnace or other purposes. In some older plants, part or all of the purge gas was vented to the atmosphere where it became a pollutant.

As Figure 4.1 shows, plants are designed to exchange heat between process streams which need to be heated and cooled respectively. Finally, it is important to note that the high pressure compressors for the feed gas are traditionally considered to "belong to" the synthesis part of an ammonia plant.

4.3 AN OVERVIEW OF HYDROGEN PRODUCTION

The major distinguishing factors among the many processes for ammonia production that have been used since 1913 are the methods and feedstocks for producing the necessary hydrogen. It is ironic that most of the processes are based on carbon (coal or coke) or hydrocarbons (natural gas, naphtha, fuel oil) as feedstocks, since the carbon ultimately fails to become part of the product ammonia, NH_3. A few processes have used pure or nearly pure hydrogen as feedstock. This hydrogen has been obtained by electrolysis of water, as hydrogen-rich refinery or chemical plant waste gases, as coke oven gas, or from the steam-iron and methane cracking reactions.

In a simplified schematic model, we can consider the production of hydrogen from a carbonaceous material, water, oxygen, and energy to proceed by a series of four steps: 1) synthesis gas production, 2) water-gas shift, 3) CO_2 removal, and 4) CO and CO_2 clean-up. (See Figure 4.2.) Unlike the ammonia synthesis reaction, the chemical species and equilibria involved in hydrogen production can be quite complex compared to the simple picture discussed here. Furthermore, the process used depends on many factors including the exact nature of the feedstock carbonaceous material as well as the general state of technical knowledge and the relative costs of feedstock, fuel, capital, and catalysts.

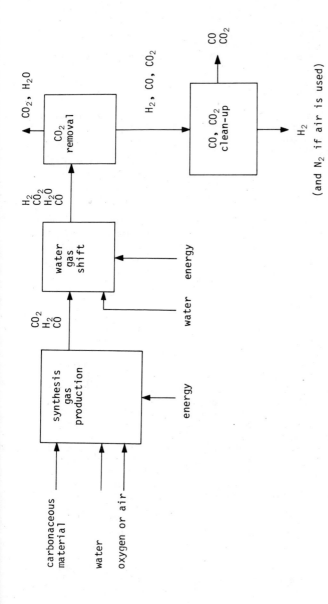

FIGURE 4.2. Schematic of Hydrogen Production From Carbonaceous Feedstock

81

In the remainder of this section we present an overview of chemistry and technology for each of the four steps. In subsequent sections we discuss the details of each of the technologies which have been or are being used for commercial production of synthetic ammonia in the United States.

Synthesis Gas Production

Synthesis gas is a generic term for a broad range of mixtures of H_2, CO, CO_2, H_2O, N_2, and hydrocarbons such as CH_4 or C_2H_6. Synthesis gas results from chemical reaction upon heating of oxygen or air, water, and a carbonaceous material. The reaction takes place, usually in the presence of a catalyst, with insufficient oxygen or air to achieve complete combustion of H_2 and C to CO_2 and H_2O.

Steam reforming of hydrocarbons such as methane or naphtha involves heating the hydrocarbon and steam in the presence of a nickel catalyst. The principle reaction that occurs is:

$$C_nH_m + nH_2O \rightleftharpoons nCO + (m/2 + n)H_2$$

This equilibrium reaction requires heat addition and results in an increase in volume of the reactants. Thus, production of CO and H_2 is favored at high temperatures and low pressures. In commercial practice, reforming takes place in two steps, with addition of air in the second stage, as discussed in Section 4.4.

Heavier hydrocarbons such as fuel oil or coal cannot easily be steam reformed. To handle them, a process called *partial oxidation* has been developed, in which the hydrocarbon is reacted with pure oxygen, or in some cases air. The oxygen is too low to produce complete combusion, and the result is production primarily of H_2 and CO with lesser amounts of CO_2 and H_2O.

$$C_nH_m + n/2\ O_2 \rightleftharpoons nCO + m/2\ H_2$$

The *coke-water gas* method of synthesis gas production dominated the ammonia field until the 1940s. In this process, coke, which is nearly pure carbon, reacts with steam as follows:

$$C + H_2O \rightleftharpoons CO + H_2$$

This equilibrium reaction requires energy input as heat, and since volume increases upon reaction, production of CO and H_2 is favored by high temperature and low pressure. Many other side reactions can occur, and careful attention to operating conditions is necessary to produce a clean CO/H_2 mixture. Coal can be similarly employed in a *coal-water gas* process. The principal difference is that volatile hydrocarbons are produced when the coal is heated, and these are reintroduced to the reactor in such a way that they are cracked to produce H_2 and CO as well.

Water-Gas Shift Reaction

In each of the processes discussed in the last section, we assumed for simplicity that a synthesis gas mixture of CO and H_2 is produced. In all cases, this mixture then undergoes the "water gas shift" reaction with steam over an iron catalyst, invented by W. Wild of BASF in 1915 [11]:

$$CO + H_2 + H_2O \rightleftharpoons CO_2 + 2H_2,$$

or neglecting the H_2 input:

$$CO + H_2O \rightleftharpoons CO_2 + H_2.$$

One can view this reaction as a process by which carbon monoxide strips a water molecule of its oxygen to free a hydrogen molecule, H_2. Note that each carbon atom thus "frees-up" one additional hydrogen molecule, which did not originate as hydrocarbon hydrogen. There is no volume change upon reaction, and considerable energy is released as heat. As a result, the equilibrium is favored by low temperatures and is relatively insensitive to pressure. On the other hand, the rate of reaction is favored by higher temperatures and by appropriate choice of catalyst. Catalyst activity is favored by high pressure.

To achieve an economic balance in energy use, equipment size, and process coordination, two stage shifting is often employed. The first stage occurs at high temperature over an iron oxide/ chromium oxide catalyst. Following intercooling, the second stage shift takes place at lower temperatures over a catalyst of mixed oxides of zinc, copper, and aluminum.

CO_2 Removal

In actual production, the output of the water gas shift reactors contains a number of impurities such as CO, CH_4, and argon as well as large quantitites of CO_2 and N_2. (The N_2 is introduced as described below.) Both the CO_2 and CO must be removed; the latter especially because it is a poison for the ammonia synthesis catalyst. The usual procedure is first to remove the bulk of the CO_2 by absorption in water or in hot water solutions of various chemicals such as monoethanol-amine or potassium carbonate. [12] The additives greatly enhance the CO_2 absorption. Much of the water in the synthesis gas leaves in the CO_2 removal section.

CO and CO_2 Clean-Up

Subsequently, the remaining CO_2 and CO are removed by one of three methods: copper liquor scrubbing, catalytic methanation, or cryogenic scrubbing with liquid nitrogen. The latter process is used in partial oxidation plants since liquid nitrogen is available from the air separation plant used to produce pure oxygen and must be introduced as a reactant for NH_3 production in any case. The most popular method in modern steam reforming plants is the catalytic methanation of CO and CO_2 with hydrogen back to CH_4 and H_2O. While this reaction is essentially steam reforming in reverse, the cost of losing a small part of the product hydrogen is acceptable compared to the cost of operating copper liquor scrubbing devices.

4.4 COMMERCIAL PROCESSES FOR THE PRODUCTION
OF AMMONIA

In this section we describe the major processes which have been used to produce synthetic
ammonia in the United States. Major attention will
be paid to the method of hydrogen production. Less
attention will be given to the design of the
ammonia synthesis reactor itself.
The approach taken in this chapter is des-
criptive and static, in order to lay a firm ground-
work for the discussion of the dynamics of innovation in later chapters. The historical use of
each process is presented in Figure 6.2 of Section
6.2 and Figures A.5 and A.6 in the Appendix. The
processes to be described in this chapter are:

1. coke-water gas process/coal-water
    gas process
2. coke oven gas process
3. partial oxidation of hydrocarbons
4. steam reforming of natural gas and naphtha
5. miscellaneous processes based on hydrogen
    recovery from chemical plants, refineries,
    and waste water.

For each process we describe the technology
used and indicate how some of the major process
parameters affect plant operation. A convenient
(although somewhat dated) summary of the effects
of pressure and temperature on plant operation are
shown in Table 4.3 taken from a book by Sauchelli
published in 1964. [13]

Coke-Water Gas Process

All of the early Haber-Bosch process plants
for ammonia synthesis were based on coke-water gas
or coal-water gas processes for hydrogen production.
Many years before large scale ammonia synthesis,
a synthesis gas consisting of CO and H_2 had been
produced and sold by municipal gas works as "town
gas" or "manufactured gas", with coal or coke as
the major raw material. The first German Haber-
Bosch plants at Oppau (Ludwigshafen) and Luena and
the first U.S. plant, Nitrate #1, were all coke-
based plants. (Nitrate #2 was a cyanamide plant.)
Synthesis gas for ammonia was first produced
from coke in a cyclic process. Air was first
blown through a bed of coke. Oxidation of a portion
of the coke made "producer gas" containing nitrogen,

85

TABLE 4.3
Effect of Temperature and Pressure on Performance of Ammonia Production Operations

Process Section	Increasing Temperature	Increasing Pressure
I. *Synthesis-gas preparation* A. Steam-hydrocarbon reforming	a) Results in low residual CH_4 and more complete reforming. b) Reduces reformer tube life c) Increases tube metal thickness. d) Increases fuel consumption. e) Increases CO content of effluent gas. f) Tends to eliminate variation of CH_4 in effluent with minor changes in feed rate and composition. g) Requires more oxygen if oxygen-enriched air used in secondary reforming step.	a) For given residual CH_4, raises operating temperature. b) For given residual CH_4 requires increase in reformer steam flow. c) Increases chances for carbon deposition; must add steam to suppress carbon formation. d) Reduces synthesis-gas compressor horsepower. e) Increases air-compression horsepower. f) Increases feed-gas compressor horsepower. g) Raises residual CH_4 for given steam and fuel rate.
or Noncatalytic partial oxidation	Where applicable essentially the same as above.	

B. Water-gas shift

a) Favors reaction rate (higher activity).
b) Increases CO leakage due to unfavorable equilibrium.
c) For given CO leakage requires more steam.

a) Increases conversion efficiency or catalyst activity.
b) Increases allowable space velocity permitting a reduction in catalyst volume.

C. Waste heat recovery

a) Increasing temperature increases heat recovery.

a) Increasing pressure increases thermal efficiency because of greater recovery of unreacted steam as latent heat.

II. CO_2 *Removal*
A. Amine system

a) Reduces solubility of CO_2.
b) Raises pressure in solvent-regeneration system making stripped CO_2 product available at higher pressure.
c) Aids regeneration of solvent.
d) Increases chances of regenerator corrosion.

a) Increases CO_2 driving force and reduces number of vapor-liquid contacts in CO_2 absorber.
b) Reduces solvent circulation.
c) Increases temperature of CO_2 regeneration system and reduces recovery of available heat in CO shift conversion effluent system (if direct exchange is used).
d) Reduces residual CO_2 in purified synthesis gas.

TABLE 4.3 (continued)
Effect of Temperature and Pressure on Performance of Ammonia Production Operations

Process Section	Increasing Temperature	Increasing Pressure
B. Carbonate systems	a) Increases absorption coefficient but also increases equilibrium CO_2 pressure above solution. b) Increases stripping coefficient. c) Increases equilibrium CO_2 pressure in regeneration system giving greater mean driving force for CO_2 stripping. d) Raises pressure in solvent regeneration system and reduces recovery of available heat in CO shift conversion effluent system.	a) Same as amine system
III. *CO Removal* A. Catalytic methanation	a) Above normal temperature range of 500-750°F, the water gas shift reaction takes place to a small extent and results in production of	a) Increases allowable space velocity permitting reduction in catalyst volume. b) For given catalyst volume improves methane conversion.

88

carbon oxides; must operate in required temperature range for practically complete conversion of carbon oxides.

B. Copper liquor scrubbing (acetate system)

a) Decreases pick-up of CO (and residual CO_2).
b) Increases NH_3 losses in absorber and stripper.
c) Improves release of CO in regeneration system.
d) Increases chemical consumption.
e) Increases lean solution cooling requirements.
f) Decreases CO_2 content of regenerated solution.

a) Increases CO and CO_2 partial pressure improving pick-up in absorber.
b) Increases high pressure pump costs.
c) Increases heat requirements for expelling CO_2 in regeneration system.
d) Promotes the reduction of Cu^{++} to Cu^+ (the active copper ion) since it allows CO to be retained in the solution long enough to effect the reduction of Cu^{++} to Cu^+.

IV. *Ammonia synthesis*

a) Favors reaction rate.
b) Decreases the equilibrium ammonia concentration.
c) Decreases catalyst life.
d) Promotes hydrogen and nitrogen attack of converter internals.

a) Favors the equilibrium ammonia concentration.
b) Permits operating at high space velocity with reduction in catalyst volume.
c) Requires higher synthesis-gas compressor horsepower.

TABLE 4.3 (continued)
Effect of Temperature and Pressure on Performance of Ammonia Production Operations

Process Section	Increasing Temperature	Increasing Pressure
IV. *Ammonia Synthesis* *(continued)*	e) Increases cooling (or refrigeration) required for product condensation. f) Requires increase in gas circulation.	d) Facilitates condensation of product ammonia in converter effluent circuit due to increased hydrogen conversion efficiency and reaction rate; if pressure is high enough, condensation of product accomplished with water cooling only.

Source: Adapted from Sauchelli [13], pp. 84-85 by courtesy of American Chemical Society.

hydrogen, CO, and CO_2. Next, steam was admitted to the bottom of the hot coke bed. As the steam was heated by the hot coke, it reacted with the coke in the upper part of the bed to produce synthesis gas or what was then called "blue gas":

$$C + H_2O \rightleftarrows CO + H_2$$

However, both heating the steam and providing energy to the reaction soon served to cool the bed below a useful temperature for reaction, usually in a matter of a few minutes. More air was then admitted, and the cycle was repeated.

The producer gas and blue gas were mixed prior to water-gas shift and gas clean up. However, the mixture was deficient in nitrogen for ammonia synthesis, and the deficit was made up by N_2 from an air separation plant. [14] Subsequently, the mixed water-gas was shifted with steam to produce a mixture of H_2, CO_2, N_2 and small amounts of CO and H_2O. This mixture was cleaned to produce a mixture of H_2 and N_2 which was then introduced into the ammonia synthesis chamber discussed in Sections 4.2 and 4.3.

The advantages of this early process for ammonia production were that it was based on well known technology and that it produced relatively inexpensive (for the times) hydrogen in large amounts. The disadvantages included the fact that the coke, which was of the high grade metallurgical variety, was used inefficiently both as fuel and as feedstock due to the losses in the cyclic process. The cyclic operation resulted in a poor load factor on each piece of equipment, since many pieces of equipment were in use during only a portion of each cycle. Furthermore, the process was dirty. To quote P. W. Reynolds of Imperial Chemical Industries Ltd. on the occasion of the substitution of naphtha reforming for coke-water gas technology:

. . . the elimination of coke put an end to the beautiful palls of violet, green, and brown smoke which rose up when the ovens were charged. It ended the black smoke and dust when the ovens were pushed and the steam and grit when the coke was quenched. Getting rid of all that, as well as the coke breeze from the gas plant and the smell from sulfur

91

removal, made a tremendous improvement in the
atmosphere of our area. The steam reformers
have four big stacks, but no one has seen
anything coming out of them except a wisp of
steam during start-up. [15]

Or to quote from F. Ernst of the government's Fixed
Nitrogen Research Laboratory in regard to a plant
for production of ammonia, sulfuric acid, and
ammonium sulfate in 1925:

A plant of this nature would have to be
located outside of town, where acid fumes and
the like would not be objectionable. [16]

Since it seemed desirable to be able to
operate coke or coal based plants continuously
rather than cyclically, in 1921 Winkler invented a
process for continuous gasification* of lignite
(low grade coal) using injection of steam and
oxygen. [17] This process was introduced on a
large scale at the Luena plant in 1930. In order
to make this process practical, large amounts of
oxygen were required. The development of the
Linde-Frankel process for air separation by
liquefaction was very important in this regard.
The liquid nitrogen produced simultaneously was
used to scrub out residual CO, CO_2, and hydrocarbons from the H_2/N_2 mixture prior to ammonia
synthesis.**

---

*The ammonia synthesis process is closely related
historically and technologically to coal gasification and liquefaction, methanol synthesis,
hydrodesulfurization, and a number of other processes.

**Many variants of the coal or coke gasification
processes were developed, and indeed are currently
undergoing much additional scrutiny as sources of
synthetic gas for utility service.
It is curious that the U.S. is now researching
means for production of methane from coal-based
synthesis gas, while many chemical processes, including ammonia, produce synthesis gas from methane
by steam reforming. At first blush, it is even
more curious that some look forward to a time when
ammonia plants will use as feedstock high BTU synthetic gas (methane) made from coal and shipped in
interstate commerce through major pipelines. Due

The Winkler process can also operate on a
coal or lignite feedstock. Coal contains more
hydrocarbon volatiles than coke, however, and these
are driven off the top of the coal bed and reintroduced at the bottom to be cracked by more steam.
Product synthesis gas is extracted near the middle
of the coal bed, relatively free of volatiles, but
in need of treating for removal of sulfur compounds and other impurities.*
    A simplified flow sheet for a Winkler, con-
tinuous coke-water gas ammonia plant is shown in
Figure 4.3. Not shown are compressors for gas
streams or heat exchangers for energy supply,
steam generation, or heat recovery. Gasification
and water gas shift take place at approximately
atmospheric pressure. The water gas mixture containing CO_2 is then compressed in four or five
stages of reciprocating compressors to 100-125 atm
in order to facilitate CO_2 removal by water
scrubbing. The high pressure is necessary because
the solubility of CO_2 in water increases dramatically
at high pressure.** The gas is then recompressed
to 200 atm (some pressure having been lost in the

---

to the costs and technical losses associated with
the intermediate methanation and steam reforming,
this approach can make sense, it appears to us,
only if enormous economies of scale accrue to a
20,000-tpd coal-based SNG plant over a 1000-tpd
ammonia plant. Alternatively, it may be efficient
to utilize synthesis gas produced in coal gasification plants directly for ammonia production by
integrating clean up and ammonia synthesis facilities with coal gasification plants.

*Exxon Research and Engineering and Halder-Topsoe
have a new water-gas shift catalyst which is promoted rather than poisoned by sulfur. This catalyst may make it easier to use sulfur-bearing coal
or oil as ammonia feedstock. See Chemical and
Engineering News, June 28, 1976, p. 16.

**Pressure on the scrubber bottoms was released to
atmospheric through a water turbine to allow CO_2
escape and to recover pressure energy. The water
was then recompressed for re-use. The CO_2 plant at
Belle, W. Va. was famous because it incorporated a
600-foot mountain at this point in the process. The
water pressure was used to drive the CO_2 solution
up the hill through a pipe and the pressure

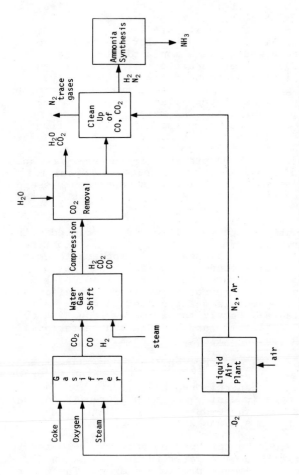

FIGURE 4.3. Schematic Continuous Coke-Water-Gas Ammonia Process

94

CO_2 scrubber) before going to a cuprous formate solution scrubbing tower for the removal of residual carbon monoxide. Following a final scrubbing with caustic solution, the gas is raised to synthesis pressure and delivered to the ammonia synthesis reactor of conventional Haber-Bosch design.

Coke-Oven Gas Process

The production of coke from coal for metallurgical purposes or as an input to ammonia synthesis is accompanied by production of a large amount of a wide variety of volatile hydrocarbons and other chemicals including SO_2 and NH_3. In fact, the nineteenth century organic chemical industry was based in large measure on "coke-oven gas" as feedstock to produce benzene, coal-tar dyes, and so on. The ammonia and sulfur dioxide that result from nitrogen and sulfur in the coal are often reacted together to produce ammonium sulfate, a solid material long used as a fertilizer.

After removal of ammonia, SO_2, benzene, coal-tars, and other heavy materials, the remaining purified coke-oven gas is rich in hydrogen (perhaps 50 percent) and methane (perhaps 25 percent). It is customary to separate this gas cryogenically by successive compression, heat exchange, and expansion. Hydrogen remains as a gas, and the other constituents, including the methane, are burned as fuel. Nitrogen from a liquid air plant is added to make the 3-to-1 H_2/N_2 mixture for ammonia synthesis by conventional Haber-Bosch techniques. The oxygen from the liquid air plant can be used as feed to a partial oxidation plant for producing additional H_2 from the mixed hydrocarbons left over from the cryogenic coke-oven gas separator, if desired. Alternatively, the oxygen may be used in a steel plant, which is likely to be nearby as the customer for the coke.*

------

was released at the top. Upon flowing back down the hill in another pipe, the water had increased in pressure nearly enough for re-injection. Thus, gravity provided a low-cost energy input.

*This is the case in the U.S. Steel coke oven gas ammonia plant built at Clairton, PA in 1969. [18]

Partial Oxidation of Hydrocarbons

As discussed in Section 4.3, nearly all hydrocarbons from light naphtha through coal can be gasified by reaction with oxygen in amounts insufficient to cause complete combustion to CO_2 and H_2O. The principal products of the complex of partial oxidation reactions are CO and H_2, which can then be shifted with steam to produce H_2 and CO_2. After CO_2 and CO removal and nitrogen addition, ammonia synthesis takes place by conventional means. The final purification of CO and CO_2 is done using the liquid nitrogen wash technique which also serves to add the necessary nitrogen to the purified synthesis gas.

The advantages of the partial oxidation process are that it is able to gasify many hydrocarbons and that a plant can be designed to use a variety of feedstocks. The disadvantage is that an expensive liquid air plant is required for the production of oxygen and nitrogen, raising overall capital investment considerably. Large volume liquid air plants have also proven to be hazardous to operate. In the 1950s, several explosions occured as a result of acetylene accumulation in liquid air facilities in ammonia plants.

Notice that modern partial oxidation plants are closely related technologically to the older second-generation coke-water gas plants based on oxygen/steam input.*

Steam Reforming of Natural Gas and Naphtha**

Process Description. In this section we discuss the processes based on steam reforming of natural gas. For our purposes here, steam reforming of naphtha, which has not been done in the U.S., can be considered to be identical.

We recognize three major periods in natural gas steam reforming technology, and for some

---

*Both TVA and ERDA (Energy Research and Development Administration) have explored new processes for ammonia based on partial oxidation of coal. See Chemical Engineering, June 6, 1977, p. 69.

**See the companion thesis by Helscher mentioned in the Preface for an extensive description and analysis of technical change in the steam reforming process over the last twenty-five years.

purposes we regard each period as representative of a separate technology. The first period is characterized by low pressure (25 psi) reforming, reciprocating compressors, and plants below 400-tpd capacity. The second is characterized by a switch to medium and high pressure reforming, up to 600 psi. The third period is characterized by high pressure reforming, centrifugal compressors, and single-train plants above 600-tpd capacity.

Since World War II, the vast majority of NH_3 plants constructed in the United States have been based on steam-reforming of natural gas. The availability of low cost natural gas, coupled with its high yield of hydrogen and the ease of processing a gaseous feedstock, has secured a dominant role for steam-reforming in the American ammonia manufacturing industry. While processes employing alternate feedstocks (coal, coke, liquid hydrocarbons, or refinery off-gases, for example) are common in other parts of the world, gas steam-reforming today accounts for 90 percent of the U.S. synthetic ammonia capacity.

Ammonia manufacture by steam-reforming natural gas is a three stage process as shown in Figure 4.4. The first of these steps involves the generation of synthesis gas containing hydrogen and nitrogen from the raw material feedstocks -- natural gas, steam, and air. Purification of the synthesis gas follows. This is a crucial step, for the ammonia synthesis catlyst is quite sensitive to the by-products formed during the generation of the synthesis gas. The final stage of manufacture involves a catalytic reaction of the purified H_2/N_2 mixture to form ammonia.

The purpose of the synthesis gas preparation stage in the steam reforming process is to produce hydrogen and supply nitrogen in proportions appropriate for ammonia synthesis. In the primary reformer, steam and most of the natural gas feedstock are catalytically converted to hydrogen, carbon monoxide (CO) and carbon dioxide (CO_2). Secondary reforming serves to react the natural gas not transformed in the primary reformer. To the process gas entering the secondary reformer is added air in an amount required to provide the proper 3-to-1 hydrogen/nitrogen ratio for NH_3 synthesis. The final step of synthesis gas production involves the conversion, using the catalytic water gas shift reaction, of by-product carbon monoxide and residual steam to additional hydrogen and carbon dioxide. This operation completes the preparation of the synthesis gas.

97

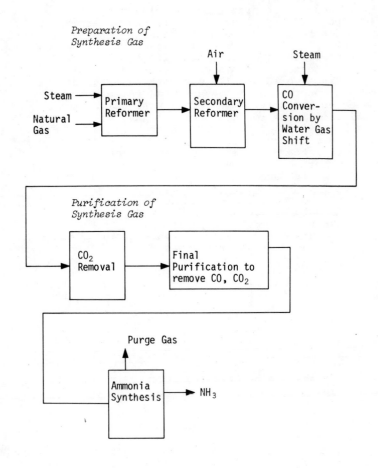

FIGURE 4.4. Schematic Steam Reforming
Ammonia Process

The gas at this point, however, contains a considerable amount of carbon dioxide, as well as small concentrations of natural gas and carbon monoxide. Protection of the ammonia synthesis catalyst requires the removal of these impurities, which is done using the techniques discussed in Section 4.3.

Improvements in Steam Reforming.* Refinements of the steam-reforming process over the last thirty years have resulted in reductions in both investment and operating costs. Advances in ammonia technology have been facilitated by improvements in the various unit operations, increases in the scale of production, and overall engineering design integration. Incremental improvements in several of the processing steps during the 1950s and early 1960s, coupled with a trend toward integrating heat and energy requirements, resulted in the appearance of a new breed of modern ammonia plants in the mid 1960s, marked by energy efficient operation and greatly increased capacity.

In traditional ammonia plants, the stages of synthesis gas preparation, purification, and ammonia synthesis were treated as a series of unrelated unit operations. Compression steps between the various processes were necessary, and little attention was given to optimizing heat recovery. Primary reforming, secondary reforming, and CO conversion were all carried out near atmospheric pressure. Carbon dioxide was removed, following a compression step, by countercurrent scrubbing with water. Another compression step was necessary before final purfication, which involved washing the synthesis gas with a regenerative ammoniacal copper solution. The synthesis of ammonia was carried out at a pressure of about 5000 psi, requiring still another compression stage. The compressors used were of a reciprocating type, which, while suited for high pressure operation, required frequent maintenance. They were driven by electric motors.

It was known that the volume of synthesis gas was much greater than the volume of feedstock gases, and that compressing the feedstock gas prior to reforming would result in significant capital and operating cost savings for compression. Furthermore, natural gas feedstock was usually

---

*This topic is also considered in Chapter 6.

available at elevated pressure from the supplier. However, materials, catalysts, and mechanical designs suitable for high pressure reforming were not available. Development of reliable reformer tubes and catalysts was underway, however, and in 1953 the first high pressure reforming ammonia plant went on stream. [19] By 1960, reforming sections of NH_3 installations were operating at pressures of 300 psi. Increased opportunity for process heat recovery came with the higher temperatures required for high pressure reforming to achieve good conversion.

During the 1950s, the methods of synthesis gas purification (both carbon dioxide removal and final purification) common in the late 1940s were succeeded by new techniques. CO_2 removal by water washing was replaced by a process of scrubbing with a monoethanolamine solution -- a system requiring less pumping capacity, able to operate efficiently at a lower pressure, and having a greater capacity for carbon dioxide removal. In the late 1950s, a catalytic method of final purification began to supplant the system of scrubbing the synthesis gas with a copper solution. The new process, methanation, offered simple and maintenance-free operation, and none of the corrosion problems which plagued the older absorptive procedures.

Ammonia plant compressors in 1960 still relied on reciprocating machines. However, many of the intermediate compression stages had been removed. A two step scheme was common: compression of the feedstock gases before reforming, then compression of the purified synthesis gas prior to ammonia formation. The process of ammonia synthesis itself had not significantly changed. Catalyst reliability had been improved somewhat, but operating pressures about 4500 psi were still typical.

Toward the end of the 1950s and into the 1960s demand for ammonia was pressing the industry's capacity. The common size of NH_3 plants in the early 1940s and 1950s was on the order of 50 tpd. By 1960 plant capacities had increased to the 300-400-tpd range.*

---

*The forces which act to constrain plant size are discussed by Helscher and in Chapter 6. Levin [23], Scherer [24] and Vietorisz and Manne [25] have explored this problem.

In 1962 a low temperature carbon monoxide
conversion catalyst was introduced. [20] Two-
stage shift convertors used the new low temperature
system and the standard catalyst in series to
obtain near complete conversion of carbon monoxide
to H_2 and CO_2. Low levels of CO reduced the load
on the final purification system and allowed the
new methanation process to totally replace the
older copper solution scrubbing technique.

In 1963 the ammonia industry experienced the
start-up of a plant featuring a compression system
combining reciprocating and centrifugal compressors. [21, 22] Centrifugal compressors, long
known to the petrochemical industry, offer considerable advantages over reciprocating machines.
First, centrifugal compressors can be driven by
simple coupling with a steam turbine. Hence,
recovered high pressure steam, available in conjunction with high temperature, high pressure
reforming, can substitute for purchased fuels to
drive the compression system. Moreover, centrifugal machines require less maintenance, are more
reliable and are easier to control, all for a lower
investment and operating cost. Discouraging the
installation of centrifugal equipment is a limit
on the maximum attainable pressure, placing an
upper bound on achievable ammonia synthesis pressure
of approximately 2200 psi. In addition, centrifugal compressors require a large minimum flow
rate.* However, with the increased demand of the
1960s, new large plants provided a gas flow rate
great enough to support a centrifugal system,
especially with reinjection of the recycled H_2 and
N_2 before the final stage of compression.

The 1963 plant combining the centrifugal and
reciprocating machines in a 600-tpd installation
introduced centrifugal compressors to modern
ammonia manufacturing. Even though the centrifugal compressors did not carry the entire compression load -- the reciprocating machine stepped
the process gas up to the final ammonia synthesis
pressure -- the successful operation of the plant

---

*A practical, physical lower limit exists to the
size of the "wheels" or rotating parts of a centrifugal compressor. Advances in compressor
design along with larger flow rates in the new
large plants combined to make centrifugal plants
competitive.

was a milestone in the technical development of ammonia manufacturing practice.

By the mid 1960s the expected cost advantages of a large ammonia plant (1000 tpd, for example) based on high pressure reforming and centrifugal compression along with the increased demand making such large plants practical, swayed the design philosophy for the ammonia synthesis section toward lower pressure operation. The lower synthesis pressures allowed the development of all-centrifugal compressor plant designs. Process studies indicated that losses resulting from the decreased conversion per pass in the ammonia synthesis section (due to the lower pressure operation) were more than offset by the gain in the overall system efficiency. Such new NH_3 facilities featured reforming pressures of up to 500 psi, highly efficient process heat recovery systems, and all-centrifugal compression systems. Reforming at elevated pressures enabled the energy recovery system to capture waste heat at temperatures and pressures high enough to power the turbine-driven compression equipment. For its role in making these innovations, the M. W. Kellogg Company was awarded the Kirkpatrick award by Chemical Engineering magazine in 1967. [27] 1965 saw the start-up of the first 1000-tpd plant and 1967 the first 1500-tpd unit. So successful has the new technology been that older, smaller plants have been forced to close, unable to remain competitive with the cost economies of large capacity installations.

The evolution of the process of steam-reforming natural gas has come about largely via chemical design and construction firms promoting their process developments among the members of the highly competitive ammonia manufacturing industry. Improved reforming technology, materials and catalysts; refined techniques of synthesis gas purification; ammonia synthesis converter and catalyst developments; and overall plant design modifications have all aided in furthering commercial NH_3 manufacturing processes. Thus, since the early 1950s, the advances in individual unit operations, the growth of markets, and the trend toward energy efficient system integration has led to the development of today's ammonia plants.*

----

*One of the latest developments in ammonia process

Miscellaneous Processes

Electrolysis. Water can be separated into H_2 and O_2 gases by passing a direct electric current between two submerged electrodes. The hydrogen so produced is usable directly, or after minor clean-up, for ammonia synthesis. Except in special circumstances, electric energy has proven too expensive for this purpose. The few plants based on electrolytic hydrogen have been adjuncts to plants for the production of chlorine by electro-lysis of sodium chloride. Proposals have been made to use cheaper, off-peak electric energy, but no such plants have been built. [28] The problem is that very large H_2 storage facilities would be needed since the ammonia plant must operate at a constant rate for technical reasons as well as to utilize its capital efficiently. Furthermore, even off-peak power is too expensive for hydrogen production today.

Steam-Iron Reaction. Very pure hydrogen can be produced by the reaction of iron and steam as follows:

$$2Fe + 3H_2O \rightarrow Fe_2O_3 + 3H_2$$

The iron oxide is then reduced back to free iron through reaction with water gas:

$$Fe_2O_3 + \text{excess } (H_2 + CO)$$

$$\rightarrow Fe + CO_2 + H_2O + (O_2 + H_2)$$

However, since the consumption of water gas is very high per unit of H_2 produced, this process has been

---

technology is the extensive application of computer control schemes for continuous optimization, fewer shutdowns, and less maintenance. The investment of a few hundred thousand dollars on computer equipment has been said to be completely recouped in a few months at todays gas prices. See *Chemical Week*, May 18, 1977, p. 43.

limited to applications in which very high purity hydrogen is required.

Refinery and Ethylene Plant Waste Gas Hydrogen. Hydrogen has been produced as a by-product of oil refinery operation and ethylene production. These hydrogen-rich off-gases have been less available in recent years, and most refineries are now consumers of hydrogen. Nevertheless, some ammonia is produced by this process. Most of the plants recover the hydrogen by cryogenic techniques and handle it in a fashion similar to purified hydrogen from coke-oven gas discussed in Section 4.4.

Thermal Cracking of Methane. Shell built an ammonia plant in 1931 using methane as a feedstock to produce hydrogen by thermal decomposition. [29] Methane is passed over a brick checkerwork which has been heated to 2200°F. It decomposes directly at those temperatures to hydrogen and carbon black, or pure carbon. The Shell plant was closed in 1967.

Refinery Waste Water Process. Chevron has designed and implemented a process for recovering ammonia from refinery waste water in two domestic refineries. [30] The process is based on steam stripping to remove the ammonia that results from destruction of organic nitrogen in the crude oil. The process reduces refinery water pollution while permitting higher levels of ammonia emissions from other refinery equipment, an overall economic advantage.

5. The Influence of Environmental Regulation on Ammonia Technology

5.1 INTRODUCTION

In this chapter we analyze the impacts of government regulation on ammonia manufacturing technology. Our approach is to identify the relevant problems of externalities, to describe the technologies available for control of those problems and their costs, and to relate the innovation of those technologies to regulatory pressures.

The emphasis in the chapter is on the modern process based on steam reforming of natural gas. A few remarks will be made on coal-based processes, and we will note also the potential impacts on technology of emerging concerns about environmental problems which may arise from application of nitrogeneous fertilizer in the farmer's field.

For each of the major environmental problems, we discuss the problem origin in an engineering sense, the various control technologies, the costs and effectiveness of those technologies, and the regulations which operate. The reader is referred to Chapter 4 for a discussion of process technologies and to Chapter 3 for detailed analyses of the regulations and other legal forces at work.

5.2 ENVIRONMENTAL PROBLEMS IN AMMONIA MANUFACTURE

We can conveniently group process-related environmental problems into three classes: water pollution, air pollution, and solid waste and land use problems. Water pollution is the most serious of these and has received the most attention, followed by air pollution. As we shall see, however, the traditional solutions to air and water

pollution have resulted in intermedia transfer
between air and water.

It should be noted that this discussion is
limited to ammonia production, and that problems
arising from production of nitric acid, ammonium
nitrate, urea, and fertilizer blends are not
included. Since, in fact, most facilities produce these products as well, treatment of ammonia
wastes is often accomplished in conjunction with
treatment of wastes from these products. Furthermore, the water and air pollution problems from
the other three products are considerably more
serious than those due to ammonia.

Water Pollution

A modern ammonia plant based on steam
reforming of natural gas produces effluent water
in five categories: process condensate, wash
solutions, cooling tower blowdown, boiler water
blowdown or steam condensate, and sanitary wastes.
The last category is handled as ordinary domestic
sewage and will not be discussed further.

Process Condensate. The primary source of
condensate water is the unreacted steam that is
removed from the synthesis gas downstream of the
carbon monoxide shift convertor and ahead of the
carbon dioxide removal system. An excess of
steam is used in the reformer and shift convertor
to ensure complete conversion of methane to
hydrogen and to guard against the formation of
free carbon in the reformer.

Small amounts of ammonia and organic compounds, principally methanol with lesser amounts
of ketones and aldehydes, are formed in the two-
stage carbon monoxide shift convertor. [1] These
compounds are dissolved in the process condensate..
In addition, the process condensate contains
lesser amounts of various materials including
sodium, iron, copper, zinc, calcium, magnesium,
and silica which originate in the catalysts,
internal refractory vessel walls, and system
piping. [2]

Differing estimates of the amounts of
impurities in the process condensate are available,
no doubt reflecting different plant designs and
operating conditions, but all based on M. W.
Kellogg technology. For a typical 1000-tpd plant,
Finneran and Welchel in 1970 [3] estimate the
process condensate to consist of 290,000 gallons/day

containing 1000 ppm ammonia and 2000 ppm organics, principally methanol. Romero and Yocum [4] report in 1975 that a 1000-tpd plant would produce about 230,000 to 288,000 gallons/day containing 1000 ppm NH_3 and 600 ppm methanol. Quartulli [5] reports in 1975 that process condensate contains 800-1000 ppm ammonia, 1000-2000 ppm methanol, 25-50 ppm other organics, 1500-2000 ppm CO_2,* and unspecified quantities of metals leached from the catalysts. These three references suggest that process condensate is produced at the rate of 1000-1200 pounds per 1000 pounds of ammonia and that it contains approximately 1.0-1.2 pounds of NH_3 and 0.8-2.4 pounds methanol per 1000 pounds ammonia. On a daily basis, a 1000-tpd plant emits about one ton of ammonia and one to two tons of methanol per day in the process condensate. Table 5.1 compares the process condensate effluent to allowable discharges of ammonium nitrogen under the Federal Water Pollution Control Act. Clearly, uncontrolled discharge of process condensate is unacceptable.

Finally, it should be noted that reciprocating compressor-based plants operate at a lower pressure in the shift reactor zone and that they can therefore be expected to have a lower level of dissolved materials in the process condensate than would a centrifugal compressor plant.

Wash Solutions and Steam Condensate. A variety of miscellaneous process waste water streams can be grouped under wash solutions. The nature of these streams depends upon the choice of CO_2 removal and final syn-gas purification techniques as well as on the kinds of air and water pollution control systems which are installed for other purposes. In a 1971 report, EPA [6] reported the composition shown in Table 5.2 for an ammonia plant waste stream including process condensate, but excluding cooling water and boiler water blowdown. We have subtracted the contribution of process condensate discussed above to estimate contributions from wash solutions, assuming that 1000 gallons of total waste water are produced per ton of ammonia as suggested in Finneran and Welchel [7].

---

*CO_2 is considered to be a significant contaminant only if the condensate is to be treated for boiler feed water.

TABLE 5.1
Comparison of Levels of Ammonia in Typical Process Condensate with Allowable
Effluent Levels

	Normalized Ammonia Levels (pounds per thousand pounds of production)	
	30-Day Average	Daily Maximum
Uncontrolled Process Condensate	1.0	1.2
New Plant Effluent	0.055	0.11
1977 Best Practicable Technology	0.0625	0.1875
1983 Best Available Technology	0.025	0.05

TABLE 5.2
Typical Ammonia Plant Waste Water Volume and Composition
(Waste Water Volume of 1000 gallons per ton of Ammonia)

Contaminant	Total Contaminants [6]		Estimated Contaminants in Wash Solutions
	(mg/l)	(lb/1000 lb NH$_3$)	(lb/1000 lb NH$_3$)
Ammonia	20 – 100	.08 – .4	0
CO$_2$	150 – 750	.6 – 3.1	1.0
Monoethanolamine	50 – 100	.2 – .4	.2 – .4
Biological Oxygen Demand	50 – 150	.2 – .4	0
Chemical Oxygen Demand	60 – 200	.25 – .84	.25 – .84
Oil	100 – 10,000	.4 – .40	.4 – .40

While the figures in the last column are order of magnitude estimates at best (especially for ammonia and BOD), they do indicate that a variety of additional chemicals are present in the wash solutions, especially MEA and oil from the compressors. According to EPA [8] MEA is a minor pollution problem, but concentrations greater than 1500 mg/l are reported to be toxic to fish upon long exposure. The high levels of oil reported here are clearly unacceptable from esthetic, human health, and environmental perspectives. In fact, oil leakage has been controlled in ammonia plants by oil separators for some time. [9] As noted in Chapter 3, EPA proposed to promulgate a maximum one-day effluent standard for oil and grease of 0.025 kilogram per metric ton of ammonia, but did not include it as part of the final standard.

The data in this section are based on a plant using MEA for CO_2 removal and methanation for final purification. Clearly somewhat different effluents can be expected if promoted hot potassium carbonate solutions are used for CO_2 removal. In addition, final purification by ammoniacal copper liquor scrubbing leads to additional release of ammonia upon regeneration of the solution to release CO and CO_2. This ammonia can, in turn, be vented to the atmosphere, be absorbed in water where it again poses a water pollution problem, or be returned to the reformer furnace as fuel supplement. Pylant [10] suggests that methanation is preferable to copper liquor scrubbing both for cost advantages and for elimination of possible water pollution problems.

Cooling Tower and Boiler Blowdown. Large quantities of water are required for cooling purposes in ammonia plants even if the plant practices a high level of energy integration. EPA [11] reports that 20,000-185,000 gallons per ton of ammonia must be circulated through a plant. When cooling towers are used, water withdrawals are a small percentage of this amount, and discharges of water, or "blowdown" are 1-2 percent of this amount. Blowdown must be employed to avoid build-up of contaminants in the circulating water which enter with the make-up water. In general these contaminants do not evaporate with the lost water in the tower. Additional contamination occurs because of in-plant leakage, and because chemicals are added in-to the cooling water to prevent corrosion, scaling, and plant growth.

110

EPA [12] reports the contaminant concentrations in
cooling water blowdown shown in Table 5.3, which
are converted in the third column to total daily
effluents based on 1000 gallons of blowdown per
ton of ammonia. Notice that ammonia effluent
from this source alone may exceed federal standards.

Little information is available on boiler
feed water blowdown, but it can be expected to
contain high concentrations of natural water background contaminants in addition to dissolved
solids, and corrosion and scale inhibitors. [13]

Air Pollution

In general, air pollution problems in modern
ammonia production are much less severe than the
water problem, and they are trivial in comparison
to those encountered in the production of nitric
acid and ammonium nitrate. Since the latter two
are often produced on the same site as ammonia,
discussions of nitrogen fertilizer complex air
pollution in the literature frequently fail to
mention problems at the ammonia complex altogether,
or mention them only in passing. (See [14] for
example.) Furthermore, the degree of air pollution
depends significantly on the adoption of systems
for water pollution control, as we shall see.
Finally, several of the potential air pollution
problems lend themselves to relatively simple
treatment by process integration.

The first air emissions in the ammonia man-
facturing sequence arise from combustion of fuel
in the reformer furnace. When natural gas is
burned the only significant air pollutant is NO_x,
mixed oxides of nitrogen, as is the case in any
industrial gas furnace. To our knowledge, no
action to control such emissions is taken or con-
templated. However, as such fuels as heavy fuel
oil, naphtha, or coal are adopted for the
reformer, we can expect to encounter pollution by
sulfur compounds, heavy metals, and particulates,
which will require controls similar to those on
electric power plants.*

_____

*A 1000-tpd ammonia plant reforming furnace is
roughly equivalent to a fifty megawatt electric
power generation facility -- not large by today's
standards but still a very significant fuel user.
The pre-WW II technology based on coke or

TABLE 5.3
Composition of Ammonia Plant Cooling Water Blowdown

Contaminant	Concentration [12] (mg/l)	Effluent Levels* (lb/1000 lb NH$_3$)
chromate	0 - 300	0 - 1.25
phosphate	0 - 50	0 - .21
zinc	0.- 30	0 - .13
heavy metals	0 - 60	0 - .26
fluorides	0 - 10	0 - .04
biocides	0 - 200	0 - .84
miscellaneous organics	0 - 100	0 - .42
NH$_3$	10 - 100	.04 - .42
MEA	0 - 10	0 - .04
sulfate	500 - 5000	2.1 - 21
TDS**	500 - 10,000	2.1 - 42
BOD	10 - 300	.04 - 1.3
COD	15 - 400	.06 - 1.7
oil	10 - 1000	.04 - 4.2

*Based on uncontrolled blowdown effluent of 1000 gallons per ton of ammonia.

**Total dissolved solids.

The second air pollution problem may arise from attempts to control water contamination by the process condensate. Alternative technologies for this purpose are discussed extensively in Section 5.3; one of the most popular of these is the process of condensate stripping. The process condensate, rich in ammonia and methanol, is mixed with steam or hot air in a large separation tower. The overhead effluent consists of a mixture of steam, air, CO_2, ammonia, and methanol; and the bottom effluent contains the majority of the process condensate water, dissolved solids, and metals. If the overhead gaseous stream is vented to the atmosphere, pollution by ammonia and methanol (a hydrocarbon) obviously occurs. More recent approaches have featured return of the overhead vapors to the reforming furnace where they are burned as fuel and add only to the NO_x and CO_2 emissions.

Another potential source of air pollution is the purge gas which must be removed from the synthesis gas recycle loop to avoid build-up of inerts and contaminants. The purge gas, which consists of ammonia, H_2, N_2, and small amounts of methane, argon, and water, has traditionally been subjected to further clean-up in order to recover the valuable components. More recently, such purge gas has been returned to the reformer as fuel, or has been used as a fuel in the ammonia oxidation unit portion of a companion nitric acid plant.

From a longer range point of view, the large quantities of CO_2 emitted by the CO_2 removal system could be viewed as a threat to the environment. However, it is now common practice to recover that CO_2 and to combine it with a portion of the ammonia product in a companion plant to produce urea, a solid nitrogeneous fertilizer. In some early ammonia plants, the CO_2 by-product was solidified at low temperature to produce "dry ice." Finally, a small amount of

---

coal-water gas production of hydrogen was a large producer of air pollutants. As early as the 1920s the importance of locating plants away from population centers for this reason was recognized.

In addition, the shift from coal to naphtha as a feedstock at ICI's Billingham, England plant was viewed as favorable in part because of the reduction in pollution. See Chapter 4.

air-borne ammonia can be expected to escape from any plant as a result of equipment leaks, spills, and opening of equipment.

Land Use and Solids Disposal

Modern 1000-tpd ammonia plants are relatively compact units which require a total land area on the order of tens of acres. In contrast, coke or coal-based plants required several times the land area for coal storage and handling, ash disposal, intermediate synthesis gas storage, complex piping and interstage cooling, and so on.

A modern ammonia plant contains several hundred tons of various catalysts which must be replaced or regenerated periodically. We did not discover any discussions of current practice or problems which might be encountered in the disposal of this material.

5.3 IMPACTS OF ENVIRONMENTAL REGULATION ON AMMONIA PROCESS TECHNOLOGY

Direct Impacts

We do not see any compelling evidence that environmental regulations have directly caused significant changes in ammonia process technology. Within the context of overall plant investment and operating costs, outlays for environmental control are small, and technologies for control have been relatively straightforward to the present.

It is customary to divide technological responses to environmental constraints into two classes: process modification and end-of-pipe treatment. However, it may be more constructive to think of the response of the ammonia industry as "process integration" in the sense that many former waste streams can be re-introduced to the process, perhaps after intermediate separation. The case of process condensate steam stripping is a case in point. A badly polluted waste water stream is separated into a gas stream containing ammonia, methanol, other organics, and CO_2 and a liquid stream containing heavy metals and suspended solids. The gaseous stream is returned to the reboiler and used as fuel; the liquid stream can be demineralized in an ion exchange unit and used as boiler feed water. The heavy metal

114

fraction from the ion exchange unit is periodically disposed of in a way that we have not ascertained. Such "process integration" is similar to the practice of exhaust gas recirculation in which auto exhaust is returned to the intake manifold; a considerable portion of the gross pollution from the process is halted but an important residual remains. It is not clear these residuals from ammonia production can be entirely eliminated, or whether they will be regulated in the near future.

Indirect Impacts

We can speculate on the indirect impacts of environmental regulation, even though we cannot identify them. One potential indirect impact is diversion of innovative activity from process innovation in response to regulation. There are three reasons to doubt that this has been the case. First, much of the innovation in ammonia processes in recent years has been carried out by large chemical construction companies, but these companies have not generally been involved in design and construction of the end-of-pipe effluent control systems which have been adopted. [15] Thus, little diversion of their efforts is likely to have occurred. Second, the pace of plant construction in the U.S. has slowed considerably in the last five years in response to difficulties with feedstock availability, price controls, and depressed markets. Thus, expansion, an important driving force for inno-vation, has been absent. Third, current trends suggest that innovative resources have been turned toward finding ways to use alternative fuels and feedstocks in the face of natural gas and liquid fuel shortages or biological and enzymatic fixation. Diversion of innovative resources to meet regulations has likely been overwhelmed by diversion to finding alternative approaches to steam reforming of natural gas. See Section 5.5 for evidence on this point.

Technologies for Meeting Environmental
Regulations: Description and Costs

Water Pollution Control Under FWPCA. A number
of options exist for the treatment of water-borne
wastes from ammonia plants. In 1971 the EPA [16]
reported that it was then the practice to employ
oil separation settling ponds, deep well injection,
or "disposal by dilution." EPA noted that dilution
was rapidly becoming unacceptable and that a number
of plants were beginning to engage in steam
stripping of process condensate. EPA further
noted [17] capital investment for water treatment
typically cost $200-1000 per daily ton of capacity
or between 8 and 78¢ per ton of product, and that
these costs were primarily for oil separation
and settling ponds. By comparison, manufacturing
costs in 1971 were typically $30 per ton.
By 1974 the situation had changed considerably. Congress had passed the Federal Water
Pollution Control Act of 1972, and EPA promulgated
final effluent regulations for ammonia manufacturing facilities in April 1974. An economic
analysis of proposed effluent guidelines was prepared for EPA by Development Planning and Research
Associates (DPRA) in January 1974 [18], which
included technologies and costs for meeting proposed BPT and BAT levels for both ammonia nitrogen
and oil and grease. (The oil and grease standard
was omitted from the final standard; see
Chapter 3.) It was proposed that the 1977 BPT
level for ammonia could be met by steam stripping,
and that for oil and grease by gravity separation.
The 1983 BAT levels could be met by ammonia air
stripping or by biological treatment.
Cost estimates by DRPA for the four technologies of interest are given in Table 5.4.
Notice that the most expensive combination of
alternatives costs were only 93¢/ton at a time
when ammonia was selling for well over $100/ton.
These costs, then, would seem to provide only a
slight incentive to innovate, not only because
they are low but also becuase they do not derive
from any technical requirement which is not
easily met by a combination of process integration
and end-of-pipe treatment. We are aware that the
DPRA estimates have been challenged, but we
believe our conclusions regarding impact on
innovation to be sound even if DPRA is too low
by a factor of 2 or 3.

TABLE 5.4
DPRA Estimates for Water Pollution Control Costs
(Basis: 340-350,000 tpy or 1000 tpd)

	Process			
	Steam Stripping	Oil Separation	Air Stripping	Biological Treatment
Investment (thousand dollars)	234	22	104	118
Annual Costs (thousand dollars)				
Energy	212	6.0	6	13
Operation and Maintenance	9	0.9	4	5
Depreciation (10%)	23	2.2	10	11.8
Interest (7.5%)	9	.8	3.9	4.4
Total	253	9.9	23.9	53.2*
Cost per ton ($)	0.73	0.03	0.07	0.15

*Includes $20,000 for extra labor.

117

Other Approaches to Water Pollution Control.
U.S. industry has apparently not suffered for want
of possible technical options for water pollution
control, although not all of them could meet
present requirements. Quartulli [19] describes
the evolution of Kellogg's treatment of process
condensate through a sequence of eight concepts
of increasing complexity and ability to control
both air and water pollution. The two most complex
of these processes are claimed to give virtually
no discharge to the environment other than combustion products, although what happens to the
trace metals and suspended solids is not made
clear.

Romero and Yocum [20] discuss eight methods
of removal of ammonia from process condensate,
six of which are different from Quartulli's eight
concepts. These methods include air stripping,
nitrification by biological reactions, ion
exchange, chlorination and charcoal absorption,
absorption by vanadium pentoxide, and precipitation
of magnesium ammonium phosphate. While some problems remain with these approaches, it is clear
that industry has many technical options from
which to choose.

Calloway, Schwartz, and Thompson [21] present
a linear programming model of selection among
various process and environmental control options
for ammonia production. However, they chose to
investigate two environmental control issues which
have not been the focus of major regulatory and
industrial concern in this industry: cooling water
thermal pollution and ultimate disposal of concentrated water-borne wastes following steam
stripping. Their results indicate that the cost
of producing ammonia and the technology employed
are very insensitive both to the cost of input
water and to disposal costs.

Air Pollution Control. We were not able to
discover any discussions of the nature or costs of
air pollution control for ammonia. Again, most of
the available data refer to control costs in
nitric acid* and ammonium nitrate plants, and one
cannot attribute all or even a significant part of
fertilizer industry total expenditures to ammonia
plants.

---

*See, for example, L. J. Ricci, "Nixing NO_x,"
Chemical Engineering, April 25, 1977, p. 70.

5.4 ENVIRONMENTAL PROBLEMS IN THE USE OF AMMONIA AND RELATED NITROGENOUS FERTILIZERS

While such problems are beyond the scope of this research, it is interesting to note that widespread use of nitrogenous fertilizers has been associated with both air and water pollution problems which may cast doubt on the long term viability of their use in current form. Living plants can only make use of fixed or available nitrogen in the form of nitrate, nitrite, or ammonia. Certain bacteria are exceptions to this rule; they can produce ammonia directly from atmospheric nitrogen and hydrogen from glucose. [22] The principal sources of fixed nitrogen for world agriculture are bacterial fixation, synthetic ammonia fertilizer, naturally-occuring nitrates, and electrostatic oxidation of nitrogen in the air in lightning discharges.*

Commoner and co-workers [24] and many others have recently investigated the fate of nitrate ion leached from the soil into water courses following nitrogen fertilizer application. While early results were greeted skeptically, it is now widely accepted that a large fraction of nitrogenous fertilizer applied according to current practice is released to surrounding water as nitrate ion or to the air as oxides of nitrogen, principally nitrous oxide.

It has proven difficult to establish any connection between high water-borne nitrate and human health problems. Nevertheless, the U.S. Public Health Service has established maximum allowable concentrations in drinking water, which some have argued are set too high, and such levels are occasionally reached in areas of high farming activity. Recent research suggests that more carefully managed timing and rates of fertilizer application can ameliorate the run-off problem. The result of such control is likely to be a reduced growth in demand for fertilizer.

More serious, perhaps, is the theory that atmospheric nitrous oxide may combine with ozone to produce nitric oxide, thus depleting the ozone

---

*Prior to the Haber-Bosch synthesis, limited amounts of nitric acid, but not ammonia, were produced commercially by a controlled high voltage arc process. [23]

layer in a manner similar to that postulated for
fluorocarbons or supersonic transport exhaust. [25]
While much remains to be learned about this theory,
one pessimistic forecast is for a 30 percent
depletion of the ozone layer if nitrogen fertilizer
use grows as it has for another century. This
growth rate implies a 1000-fold or greater increase
in fertilizer use, a most unlikely outcome. Other
forecasts are much less pessimistic, and most
scientists stress our limited understanding at this
time. Nevertheless, should the theory be confirmed, some very hard times are ahead for the
fertilizer industry, for agriculture, and, thus,
for all of us.

5.5 EVIDENCE FROM INDUSTRY TRENDS
FOR IMPACTS OF REGULATION

Previous sections of this chapter were based
on an analysis of what the ammonia industry has
done in response to regulatory pressures. In this
section we report on a different approach: an
analysis of what the industry said it was concerned about.
In order to document our sense of important
trends in the ammonia industry, we performed a
simple content analysis of all articles on ammonia
which appeared in the trade journal Chemical Week
in the years 1965, 1970, 1973, 1974, 1975.* We
abstracted a total of 168 articles which were
relevant to some aspect of ammonia production or
marketing. The articles ranged in length from a
single paragraph to full-length features.
Chemical Week, published by McGraw-Hill, is
oriented toward production engineers, marketing
people, and management in the chemical industry.
Its articles tend to be non-technical, but
important technological innovations are noted or
featured regularly, as are such external trends as
weather or government actions which might affect
markets. Thus, one would expect the journal to
signal primary industry concerns by the degree to
which they receive attention in its pages.
Table 5.5 shows a breakdown of the 168
articles by primary topic and year on a percentage
basis. Two shortcomings of this table must be

---

*Issues for 1976 and 1977 were unavailable at the
time the abstracting was completed.

TABLE 5.5
Principal Topics of Ammonia Fertilizer Related Articles in _Chemical Week_ 1965 – 1975

Topic	Percent of Articles in Each Year				
	1965	1970	1973	1974	1975
New Plant Construction and Investment	25	5	3	17	2
Short Range Finances and Markets					
Prices, Sales, Weather, Inventories	18	52	68	39	48
Imports, Exports, Foreign Markets	6	10	10	13	7
Acquisition, mergers, legal problems	25	5	--	--	5
Long Range Demand Forecasts	8	--	--	4	--
Environmental problems in production	--	--	--	4	--
Environmental problems in use	--	5	--	--	--
Occupational safety and health	--	--	--	--	2
Transportation of ammonia	2	5	3	--	2
Gas shortage, feedstock shift, etc.	--	10	13	13	10
Process Innovation	14	--	--	4	21
Production economics	--	10	--	4	--
Miscellaneous	2	--	3	--	2
Total Number of Articles	51	21	31	23	42

mentioned at the outset. First, each article was
assigned to only one topic area, based on a twenty-
five word abstract. For perhaps ten percent of
the articles more than one topic might have been
appropriate; where possible we assigned such
articles to the areas of interest to this study:
environment, OSHA, feedstock shifts, and process
innovation. Second, we have no way of knowing
whether the shifts in emphasis are those of the
entire industry, of a vocal portion of the industry,
of management or technical people, or of the
journal itself because of changes in editorial
policy or staff interest. Nevertheless, a number
of interesting observations can be made about
Table 5.5:

a) Only one article each dealt with the
impacts on the industry of environmental and
safety and health regulations. Such regulations
have not been major areas of concern.*

b) In 1965, the industry was in a great
state of flux. A number of new plants were
announced, the structure of the industry experienced rapid change through mergers and acquisitions,
and process change was widely discussed. Con-
versely, feedstock (natural gas) availability was
not mentioned at all.

c) By 1970, articles on new plant, on process innovation, and on industry structure had
nearly disappeared, while concern for natural gas
availability and feedstock shifts had emerged.

d) The very great attention to short term
markets in 1973 reflects two factors. First,
spring flooding in the midwest upset agricultural
use through the normal marketing period. Second,
price controls were widely debated and then
lifted in late 1973, leading to major price
increases and more discussion. Concern for feedstock availability persisted.

e) 1974 witnessed a resurgence in plant construction announcements, which did not, however,
persist into 1975.

f) Perhaps the most striking feature of
1975 is renewed interest in new processes. These
take three forms: i) incremental innovation for
cost-cutting in the face of energy, feedstock
and equipment price hikes; ii) improved catalysts;

---

*Similar analyses might be used to select more
promising products for study based on a more frequent reference to regulation.

122

and iii) entirely new concepts including ammonia
from municipal-waste-based-gas, atmospheric
pressure synthesis, biological synthesis in combination with plants, and sea-going floating production facilities.

g) The steady concern for short term financial and marketing news reflects both the
chronic dependence of markets on seasonal demand
and weather and the chronic cyclic mismatch
between productive capacity and demand for the
product.

In summary, we would highlight three observations. First the industry was in a state of
flux in 1965 and appears to moving that way again.
Second, feedstock price and availability have
become major industry concerns in recent years.
Third, through 1975 environmental and safety and
health regulations were not of major concern to
the ammonia industry.

5.6 CONCLUSIONS

The most important finding of this chapter
is that environmental regulation appears to have
had some, but only a small, impact on ammonia
manufacturing technology. A coalescense of forces
contributed to that result. The regulations are
not very significant for the industry, and cost
of compliance with those that exist is probably
relatively minor. Air emissions from ammonia
manufacturing processes are not directly regulated.
Further, air emissions from the manufacture of
other nitrogen fertilizers vastly overshadow
ammonia manufacturing emissions. EPA regulations do
directly affect water effluents at ammonia manufacturing plants. However, these standards probably do not present a major incentive to innovate
for a combination of two reasons: (1) the industry
has obtained standards which it indicates are
generally acceptable (see the discussion of
Vistron v. Train in Chapter 3), and (2) costs of
compliance with those standards are not very
large. Our review of articles in Chemical Week,
which show very little interest in environmental
regulation of nitrogen fertilizer manufacture,
reinforces our belief that such regulations have
had only a limited impact on ammonia manufacturing technology.

Nonetheless, we also observe that industry is
reacting to the environmental damage done by

123

ammonia manufacture. For example, it is quite
clear that new plants are being designed to reduce
pollution by recirculating those air emissions
which are useful as a fuel supplement back to the
boilers and by evaporating waste ammonia/water
solutions to recapture marketable, diluted ammonia.
Other examples can be found. Although present
environmental regulations are not providing the
stimulus for these activities, the attempts to
innovate may be attributed both to the anticipation
that new regulations will be imposed and the
possibility that expected future fuel and ammonia
prices will make such practices profitable.

6. Application of Production and Innovation Models to Steam Reforming of Natural Gas

Our goal in this chapter is to apply the production and innovation models presented in Chapter 2 to the production of ammonia by steam reforming of natural gas. In Section 6.2 we examine the major innovations qualitatively, noting the scientific and engineering problems which had to be overcome to make the innovations and the general effect of the innovations on inputs and capacity. This discussion is organized around the four types of innovation described in Chapter 2. Results of this part of the analysis are summarized in tables that provide values of the most important variables describing the process at various points in time. This and other published information is used to create a set of input coefficients which represent the technology embodied in newly constructed plants.

In section 6.3 these input coefficients are examined in a number of regression models to test the hypothesis that they are affected by input price changes and other factors.

Since published process data are extensively utilized in this chapter we begin in Section 6.1 with a brief discussion of the quality of those data and of their major strengths and weaknesses.

6.1 QUALITY OF AVAILABLE PROCESS DATA: FACTOR INPUTS AND INVESTMENT

In Chapter 4 we presented engineering descriptions of several processes for producing ammonia. These descriptions are useful for identifying safety or environmental problems associated with a process and for indicating opportunities for substitution between inputs. To study

effects of input price changes, however, it is
necessary to utilize data which indicate the use
of purchased inputs.

Suitable data for this purpose appear in the
engineering literature, a typical example of which
is reproduced as Table 6.1. It will be noted that
the table includes data on the following inputs,
in physical quantities: raw materials, fuel,
water for boilers and for cooling, and man hours
of labor. Catalysts and chemicals costs are presented on a cost per ton of output basis. Maintenance is given as a proportion of fixed
investment. Finally, for a specified capacity,
estimated expenditures on fixed capital in current
dollars are given. Such process coefficient
information often appears in chemical engineering
journals, books, and United Nations publications.
The authors are generally employed by companies
which design and construct plants or by government
agencies. Since this material is widely diffused
and since the same information occasionally appears
in more than one place, it is unlikely that serious
inaccuracies appear in the data. Moreover, experience with the diffusion of processes is consistent
with the claims made for the processes in the
literature. On the basis of these considerations,
it is likely that large differences in plant design
will be reflected in differences in input coefficients.

A general problem with the data is the lack
of documentation -- it is very difficult to know
what the data are based upon. It is less likely
that they are based on the actual experience of a
plant than on estimates made by plant designers.
Although we have been assured that actual experience is rather close to estimates, lack of data on
actual experience makes it difficult to observe
certain types of technical change, such as
"learning by doing" within a particular plant.*

A second major caveat is that the data are
frequently not explicit regarding the items which
are included in the fixed investment category; it
is frequently difficult to know whether such items
as storage, cooling towers, and site preparation
have been included along with the basic plant or

---

*We have been assured by several persons in the
industry that modern plants are highly optimized
and that little room exists for improving the performance of a plant in place.

TABLE 6.1
Typical Literature Data on Ammonia Production Costs

Plant Capacity (tpd)	200	300	400	600	1000	1500
Compressors & Drives	Reciprocating—Motor			Centrifugal—Steam Turbine		
Investment ($)	4,300,000	5,500,000	6,500,000	8,500,000	12,300,000	15,750,000
Raw Material & Utility Requirements						
Natural Gas (million BTU/ton) . . .	30.0	30.0	30.0	32.3	32.0	31.5
Power (kwh/ton) . . .	625	625	625	30.0	30.0	30.0
Boiler Feed Water, Make-Up (gallons)	550	550	550	550	550	550
Circulating Cooling Water (gallons) .	50,000	50,000	50,000	50,000	50,000	50,000
Production Costs ($/ton of ammonia)						
Natural Gas (@20¢/million BTU) . . .	6.00	6.00	6.00	6.46	6.40	6.30
Power (@0.7¢/Kwh) . . .	4.37	4.37	4.37	0.21	0.21	0.21
Boiler Feed Water (@25¢/1,000 gallons) .	0.14	0.14	0.14	0.14	0.14	0.14
Circulating Cooling Water (@2¢/1,000 gallons) .	1.00	1.00	1.00	1.00	1.00	1.00
Total Raw Material and Utility Costs ($/ton)	11.51	11.51	11.51	7.81	7.75	7.65
Other Costs ($/ton)						
Catalyst and Chemicals	0.40	0.40	0.40	0.40	0.40	0.40
Operating Labor (4 Men & 1 Supervisor-$3/hr)	1.80	1.20	0.80	0.60	0.36	0.24
Plant Overhead (100% of Operating Labor) . . .	1.80	1.20	0.90	0.60	0.36	0.24
Maintenance (3% of Capital Investment) . . .	1.84	1.57	1.37	1.22	1.05	0.90
Taxes and Insurance (2% of Capital Investment)	1.23	1.05	0.91	0.81	0.70	0.60
Interest (5% of Capital Investment) . .	3.07	2.62	2.28	2.02	1.76	1.50
Amortization (10% of Capital Investment) . . .	6.15	5.24	4.57	4.05	3.52	3.00
Total Production Cost ($/ton)	27.80	24.79	22.84	17.51	15.90	14.53

Source: [1], p. 197, by courtesy of Gulf Publishing Company.

127

"battery limits" costs. This is presumably less of
a problem when several plants are discussed in the
same article, or by the same authors. A related
difficulty with the capital investment is the
extent to which estimates for plants of different
capacities are made independently, rather than by
extrapolation from one capacity. The engineering
literature relies rather heavily on the "six-
tenths" rule or one of its variations*, so that our
observation of this relationship between capital
costs and capacity may merely recover the authors'
unstated assumptions. Finally, as we discussed in
Chapter 2, differences between coefficients of the
same process may arise because of different "make
or buy" decisions. A firm may choose to produce
all or part of its own electricity and will therefore be observed to purchase relatively large
amounts of natural gas. The author may fail to
state this type of substitution; in view of the
level of aggregation at which investment appears,
it may be difficult to distinguish such substitution from process change.

Thirdly, for our analysis we assume that a
set of data appearing in the literature at a point
in time are typical of data for the newest process
in use at that time. With this assumption, we can
study process coefficients for the innovative
technology rather than industry-wide average
coefficients such as are available in input-output
tables.

To summarize, although there are difficulties
with the process coefficient data as found in the
engineering literature, they are a rich source
of information about technical change at a very
low level of aggregation. Accordingly, we believe
it is worthwhile to work with them as indicators
of technical change.

6.2   INNOVATIONS IN THE PROCESS BASED ON
      STEAM REFORMING OF NATURAL GAS [2]

In this section, we use developments in the
steam reforming of natural gas process to illus-
trate three of the four types of innovation discussed in Chapter 2: reducing inputs, broadening

---

*Plant investment is often estimated to be pro-
portional to plant capacity raised to the six-
tenths power.

substitution possibilities, and increasing scale.
Examples of the fourth type, process change, may be
found in Chapter 4. The primary intent of the
section is to establish that the process developments summarized here were, in fact, innovations;
that is, they required nontrivial research and/or
development expenditures and became available for
industrial use within the time period being discussed. Whenever possible, we note the scientific
and engineering problems which were overcome to
introduce the particular development.

Input-Reducing Innovations

Advances in the catalysts for ammonia
operations are an excellent example of innovations
which reduce overall input requirements. Improvements took place in catalysts for reforming,
shift conversion, and methanation. Increased
catalyst efficiency and activity have allowed
reductions in the reactor volume needed for a given
operation and at the same time improved the degree
of conversion. The net results of catalyst
advances are smaller equipment sizes and lower fuel
and feedstock requirements. The best illustration
of how developments in catalyst technology can
reduce input coefficients is the introduction of
the low-temperature carbon monoxide conversion
catalyst. The low-temperature catalyst permitted
a reduction in capital equipment -- either the
replacement of a copper liquor system of final
purification with a methanation arrangement or
the elimination of a second stage of carbon
dioxide scrubbing in the case of two-stage $CO-CO_2$
designs. Moreover, the more complete conversion
of CO to CO_2 with the new catalyst increased
hydrogen production, and hence lowered the natural
gas input required to generate a given amount of
H_2.
It must be emphasized that the introduction of
a new catalyst is not a simple substitution.
Identifying the most appropriate catalyst for a
chemical reaction is an expensive, largely trial-
and-error, activity. 2500 substances were examined
before a successful ammonia synthesis catalyst was
discovered at BASF in the early part of this
century. Only in recent times has catalyst design
been put on a reasonably scientific basis. [3]
Moreover, there presently exist a number of firms
which specialize in R&D for catalyst improvements,
suggesting that this is a research-intensive

129

activity, not merely substitution. Moreover, such firms are likely to direct their research toward industries which appear to yield the greatest return.

The developments leading to high pressure reforming operations are also examples of innovations having the effect of reducing the inputs to the ammonia process. Raising the reforming pressure was accompanied by decreases in investment, natural gas, and utilities costs. The relevant information is presented in Table 6.2. It may be seen that all of the input requirements decline or remain constant as pressure is increased. While raising the reforming pressure falls into the first category of innovation, it also has implications for increasing capacity, as discussed below. The input reductions achieved by higher pressure operations required considerable innovative activity. We briefly summarize these developments to suggest the nature of the barriers and the type of solutions achieved.

It was recognized early that reforming at an elevated pressure might benefit ammonia manufacturing practices. The particular advantage of high pressure reforming that first appealed to designers was the possible reduction in the synthesis gas compression horsepower. The reduction derives from the fact that the volume of gas to be compressed increases as the process gas proceeds from its raw input state to the final purified synthesis gas form; hence compressing at an early stage entails compressing a smaller volume of gas. However, reformer tubes capable of withstanding the high temperature and pressure conditions of the reforming operation were not available. To overcome this obstacle, one approach investigated in 1954 was to install a larger number of reformer tubes having smaller diameters and relatively thicker walls. [4] This innovative effort to reduce the effective stress in the tube walls was an attempt to improve existing tube designs marginally, but it was a limited approach that did not allow for additional cost reductions possible in high pressure reforming, such as heat recovery.

Referring to Table 6.3 note that the compressor operating costs, reflected in the kilowatt-hours per ton of output, are reduced by about 20 percent as reforming pressure is raised to 150 psi. However, the early higher pressure operation required higher reformer temperatures and the installation of pressure-resistant equipment

130

TABLE 6.2
Effect of Reforming Pressure on Investment, Fuel, and Utility Inputs

Reforming Pressure (psi)	205	275	350	400
Estimated Investment (excluding catalyst)	$5,400,000	$5,090,000	$4,960,000	$4,890,000
Feed Gas (million BTU/ton)	20.32	20.32	20.32	20.32
Utilities				
Fuel gas (million BTU/ton)	10.26	9.86	9.64	9.40
Power (kwh/ton)	680	652	634	622
Cooling water (gallons/ton)	48480	44880	42240	40608
Boiler Feed Water (gallons/ton)	622	578	552	552

All other factors are identical through the pressure range.

131

Source: [5]

TABLE 6.3
Ammonia Production Inputs for Low and High Pressure
Reforming in a 120-tpd Plant in 1954

Inputs	Reforming Pressure (psi)	
	45	150
Investment (million dollars)	4.07	4.21
Variable Inputs per ton of NH_3		
Feed gas (million BTU)	22.3	22.3
Fuel gas (million BTU)	11.2	13.5
Total* (million BTU)	44.9	45.2
Power (Kwh)	858	676
Cooling water (gallons)	79500	77100
Catalysts and Chemicals ($)	1.84	1.81
Warehouse Labor ($)	1.80	1.80
Maintenance (% of investment)	3	3
Total manufacturing cost ($/ton)	52.20	51.68

*Includes 2-million BTU purge credit and natural gas
required for steam generation.

Source: [6]

throughout the rest of the gas preparation section, resulting in both increased capital and fuel charges. Nonetheless, there was a fall in the cost per unit of output.

Throughout the 1950s, the reforming pressure inched its way up, reaching 250 psi by 1960. Although it is not entirely clear just what improvements permitted the pressure to reach this level, it appears that advances in materials and equipment technology made important contributions. Certainly, the development of better reformer tubes and a better furnace design played a central role. The evidence indicates that costs of electricity for compression as well as natural gas inputs declined as pressure rose. Further, there is reason to believe that unit capital costs also declined. In any event, since reforming pressures did continue to rise, it must be assumed that the net effect of increasing reformer pressure on the cost of production was favorable.

Further improvement of tube materials apparently did not hold the key to removing the 300-psi limit on reforming pressure. In 1964 there remained limits on the temperature/pressure conditions under which reforming operations could reliably be carried out. The key to Kellogg's successful hurdling of this barrier was not to hurdle it at all. Rather, by modifying the design of reforming operations, tube-wall considerations no longer represented an obstacle to higher pressure reforming practices. Kellogg's system of heating the primary reformer effluent and preheating the secondary air resulted in a change in the balance of primary to secondary reformer duty. The net effect of this modification was a reduction in the temperature required in the primary reformer. The lower temperature in turn permitted higher pressure operation without violating tube-wall stress limits. Catalyst improvements further removed tube wall considerations from critical concern. Finally, higher reforming pressures permitted increased possibilities for heat recovery which reduced the unit fuel (gas) input [7], and the "reduction in investment results from reforming at higher pressures, more complete data on the reactions, improved catalysts, and improved schemes for heat recovery." [8] However, the diminishing returns to each of these factors as pressure rises signals the approach of the new capital and power restrictions on reforming pressures. Thus, in contrast to the earlier mechanical and technological

133

approaches, Kellogg reduced the impediment to higher
pressure reforming by concentrating on system
design. The improved understanding of the relationship between primary and secondary reformer
operations together with advances in reforming
catalysts shifted the barrier to higher reforming
pressures from tube wall considerations to costs
of compression.

Innovations Which Increase Substitution Possibilities

The evolution of steam-reforming techniques
offers a number of examples of process changes
which increase substitution possibilities. As an
example, consider the development of three alternative methods of carbon monoxide shifting and
removal in the mid-1950s.* Single-stage CO
shifting with a copper liquor scrubbing system for
final purification, two-stage CO shifting followed
by methanation, and single-stage CO shifting
followed by methanation were all means to accomplish the same end -- removal of carbon monoxide
from the synthesis gas stream. None of the three
practices dominates; each offers differing blends
of investment and operating requirements (see
Table 6.4), and each has its own particular set of
advantages and disadvantages. The availability of
three techniques to accomplish the same end,
however, gave ammonia plant designers the flexibility of selecting that system best suited to the
input price situation and the location of each
individual NH_3 plant; the substitution possibilities
had been expanded beyond the earlier limitation to
the single-stage CO/copper liquor design alone.
Since the mid 1960s the hot potassium method
of CO_2 scrubbing has increased the substitution
range.** In areas of low feedstock costs, designers

---

*Carbon monoxide is removed from the synthesis gas
through CO shifting and final purification. The
means of CO_2 removal is of little consequence to
this discussion and hence will be neglected.

**Although hot potassium carbonate methods of CO_2
removal had been available for ammonia duty since
the mid 1950s, it was not until reforming pressures
had climbed and the carbonate methods themselves
had been improved that they were widely accepted by
ammonia plant designers.

TABLE 6.4
Investment, Feed and Utility Costs for Various Process Flowsheets [1965 data]

Basis: (1) Ammonia Capacity 100 tpd (4) Two One-Half Capacity Compressors
(2) Natural Gas Feed at 220 psi (5) Gulf Coast Location
(3) Primary Reformer Pressure 220 psi (6) Cooling Water Temperature Rise 25°F

Case	A		D		E	
Process Sequence	Single Shift, MEA Scrubbing, Methanation, Synthesis		Single Shift, MEA and Copper Liquor Scrubbing, Synthesis		Double Shift, MEA Scrubbing, Methanation, Synthesis	
Compressor Drive	Motor	Gas Engines	Motor	Gas Engines	Motor	Gas Engines
Estimated Investment (excluding Catalyst)	$2,830,000	$3,280,000	$3,020,000	$3,470,000	$2,880,000	$3,330,000
% CH_4 in Purified Synthesis Gas	1.33	1.33	0.23	0.23	0.62	0.62
Feedstock Gas (million BTU/hr)	86.5	86.5	81.1	81.1	81.4	81.4
Utilities						
Fuel Gas (million BTU/hr)	53.6	81.5	53.8	81.5	54.4	81.7
Power (Kw)	3,090	178	3,050	236	3,050	187
Cooling Water (85°F) (gallons/minute)	3,610	3,860	3,400	3,650	3,940	4,190
Boiler Feed Water (gallons/minute)	69.0	69.0	69.3	69.3	74.3	74.3
Steam Export (pounds/hr)						
High Pressure	12,500	12,500	12,700	12,700	5,900	5,900
Low Pressure	3,000	3,000	3,600	3,600	3,400	3,400
Cost of Feed and Utilities ($/day)						
Price Structure No. 1						
Natural Gas @ $0.32/million BTU	1,076	1,290	1,036	1,249	1,043	1,253
Power @ $0.009/KWH	667	38	659	50	659	40
CW @ $0.02/thousand gallons	104	111	98	105	114	121
BFW @ $0.25/thousand gallons	25	25	25	25	27	27
Steam @ $0.50/thousand pounds (Credit)	(-)186	(-)180	(-)196	(-)196	(-)112	(-)112
Total	1,686	1,278	1,622	1,233	1,731	1,329
Cost Per Ton NH3 ($)	16.86	12.78	16.22	12.33	17.31	13.29
Price Structure No. 2						
Natural Gas @ $0.20/million BTU	672	806	648	780	652	783
Power @ $0.006/KWH	445	26	439	34	439	27
CW @ $0.02/thousand gallons	104	111	98	105	114	121
BFW @ $0.25/thousand gallons	25	25	25	25	27	27
Steam @ $0.40/thousand pounds (Credit)	(-)149	(-)149	(-)156	(-)156	(-) 89	(-) 89
Total	1,097	819	1,054	788	1,143	869
Cost Per Ton NH3	10.97	8.19	10.54	7.88	11.43	8.69

Source: [9]

may conserve on investment by installing a MEA CO_2
removal system. While reducing the capital expense,
they will, however, face higher operating charges.
In areas of high feedstock costs, the more capital
intensive potassium carbonate CO_2 removal system may
well be selected on the basis of its lower input of
natural gas. Table 6.5 summarizes the capital/
energy tradeoff situation for MEA versus carbonate
CO_2 scrubbing practices.

The development of the catalytic approach to
final purification by methanation is also an innovation that has broadened the substitution possibilities within the steam-reforming process. (See
Table 6.6.) However, while the examples concerning
CO conversion and CO_2 removal techniques involved
clear-cut capital/operating cost choices, the
selection criteria for choosing either the methanation or copper liquor method of final purification
include an additional factor -- the superior reliability and controllability of the methanation
system. Thus, the move to methanation has the
flavor of both a reduction of inputs and an increase
in substitution possibilities.

To summarize, carbon conversion and removal
subprocesses are closely interrelated, and a variety
of techniques featuring different input configu-
rations have been developed. (Details of the history of this development may be found in Helscher's
thesis.) The optimal design at any given time for
carbon oxide removal requires coordination of the
individual unit operations and consideration of input costs. Substantial innovative activity has been
undertaken since the 1940s to improve these aspects
of ammonia production. The innovations depended
upon developments in high pressure reforming and
have required research into catalysts as well. Although most of the research for these developments
took place in private firms, it is interesting to
note that the "hot" potassium carbonate method was
developed by the Bureau of Mines. [12]

Our final example of substitution possibilities
concerns heat recovery. Selecting the degree of
heat recovery is, like the choice of a CO_2 removal
technique, often associated with the trade-off between investment and feedstock costs. In areas of
low feedstock costs plant designs generally utilize
low thermal efficiencies and intermediate pressure
steam systems. In areas of high feedstock costs,
designers will incorporate high pressure steam
systems in NH_3 plant flowsheets, featuring a high
degree of waste heat recovery. As in the case with

TABLE 6.5
Input Requirements for CO_2 Scrubbing with MEA
and Carbonate Systems (1000-tpd plant)

	CO_2 Removal System	
Inputs	MEA	Hot Potassium Carbonate
Investment	$20,000,000	$20,300,000
Feed Gas (million BTU/ton)	20.40	20.40
Utilities		
Fuel Gas (million BTU/ton)	12.74	10.56
Power (KWH/ton)	15.4	15.4
Cooling Water (gallons/ton)	2448	1858
Boiler Feed Water (gallons/ton)	527	560

Source [10]

TABLE 6.6
Input Requirements for Final Purification by Methanation
and Copper Liquor Scrubbing (1000-tpd plant)

	Final Purification System	
Inputs	Methanation	Copper Liquor
Investment	$2,830,000	$3,020,000
Feed Gas (million BTU/ton)	20.76	19.46
Utilities		
Fuel Gas (million BTU/ton)	12.86	12.91
Power (KWH/ton)	742	732
Cooling Water (gallons/ton)	51,984	48,960
Boiler Feed Water (gallons/ton)	993	998

Source [11]

137

carbonate CO_2 removal practices, high pressure steam systems require a bit more capital, but save on operating inputs. Table 6.7 compares the input coefficients for high and low pressure steam systems which feature higher and lower degrees of heat recovery, respectively. The development of steam systems of differing efficiencies, and correspondingly different capital/operating input requirements, represents an expansion of the substitution possibilities facing NH_3 plant designers.

There has been a clear trend in the design of ammonia plants to increase plant operating efficiency through the reuse of energy and water. The methanation/low-temperature CO shift catalyst combination, hot carbonate CO_2 removal technique, and high efficiency heat recovery systems are all substitution possibilities which allow today's NH_3 designers to react to changing capital and energy prices.

Innovations Which Foster Scale Increases

There have been two rather distinct periods of scale expansion in steam-reforming plants. (See Figure 6.1.) Through the 1950s the size of the largest ammonia unit rose from about 200 to nearly 400 tpd. In the 1960s plant size jumped to 600, then 1000, and shortly thereafter to 1500 tpd.

Scale expansion through the 1950s was apparently fostered by the development of higher pressure reforming techniques. As reformer pressures climbed, the physical size of reforming furnaces for larger NH_3 units became manageable. Gas holding facilities were eliminated and ammonia plants became continuous-flow processes. Plants of a capacity that would have required multiple-train facilities under atmospheric reforming conditions could be designed in more economical single-train arrangements (except for the synthesis compression section). There is reason to believe that limits to market size restricted scale expansion of NH_3 production units in the 1950s. [15] No ammonia facilities much larger than 400 tpd, the upper limit that Finneran [16] cites as typical for a local distribution center, were constructed at that time. In addition to market constraints, there were technical limitations on compressor capacity to which we next turn.

The standard ammonia plant through the 1950s and into the early 1960s utilized reciprocating

138

TABLE 6.7
Input Requirements for High Pressure and Low Pressure
Steam Systems (1000-tpd plant)

Input	Steam System	
	High Pressure (1500 psi)	Low Pressure (650 psi)
Investment	$20,000,000	$19,350,000
Feed Gas (million BTU/ton)	20.40	20.40
Utilities		
Fuel Gas (million BTU/ton)	12.74	16.08
Power (KWH/ton)	15.4	15.4
Cooling Water (gallons/ton)	2448	2966
Boiler Feed Water (gallons/ton)	527	527
Degree of Heat Recovery	High	Low

Source [13]

compressors. The possibility of reaping economies
of scale within this design depended upon the
ability to increase the capacity of these compressors, but there were technical difficulties.
A 5000-horsepower compressor was the largest
reciprocating machine available in the mid-1950s.
Two compressors of approximately that size, each
carrying half the load, would be required in a
300-tpd plant. Although 10,000 horsepower compressors were available in 1964, their reliability
was suspect. [17] A 600-tpd plant requires two
operating 11,750-horsepower compressors. Large
reciprocating compressors require considerable
maintenance and are unreliable, noisy, and difficult
to control. Furthermore, nowhere in the 1950s or
early 1960s literature is the installation of a
third synthesis loop compressor, which would have
been necessary to satisfy the horsepower
requirements of a large recirpocating plant, con-
templated. Hence, it seems reasonable to conclude
that either three-compressor arrangements were

139

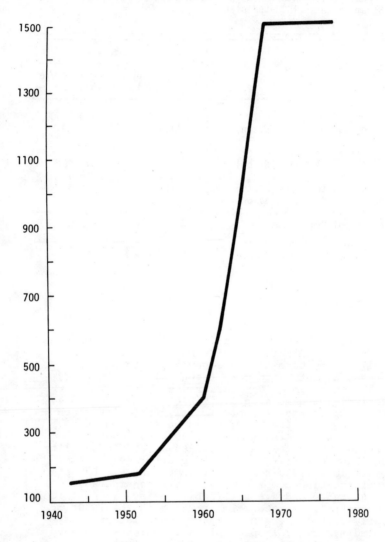

FIGURE 6.1. History of Largest U.S. Ammonia
Production Units. Source: [14]

uneconomic, or plants of a capacity requiring so
great a compressor horsepower simply were not at
that time being considered due to market limits, or
both. Of course, a third compressor, by itself,
would not have yielded significant economies of
scale. Although reciprocating designs as large as
600 tpd might have offered. lower production costs
than smaller plants of the same general configuration, such large reciprocating designs were
probably not available in the late 1950s. In fact,
none have been built.

The problem of compressor capacity was solved,
not by increasing the size of reciprocating compressors, but by switching to centrifugal compressors. These compressors require a large volume
of gas, can be run with steam already being
generated in the plant, and have desirable maintenance and noise characteristics. Since centrifugal compressors are not capable of achieving
pressures as high as those reached by the reciprocating type, process modifications were developed
to allow lower synthesis pressures. The new configuration was obviously judged to be superior by
producers: within a very short time such plants
were virtually the only type built. (See Figure
6.2.)

As noted above, in addition to technological
factors which discouraged scale increases, there
were limits to market size due to storage and
transportation costs. Having satisfied the local
demand, no method was available to transport
economically the rest of a large producer's NH_3
to distant markets. The costs of transportation
would quickly outweigh the slightly lowered production charges possible in bigger reciprocating
plants; ammonia plants were traditionally located
near their markets because of the high cost of
rail and road transportation.

Not only were distant markets relatively
inaccessible in the late 1950s, but local markets
were oversupplied as well. For the period of
1955-1959 there was a serious condition of NH_3
overcapacity in the United States. [19] In 1955
Chemical Week reported that "the gap between
projected capacities and current markets for
anhydrous ammonia is widening." [20] By 1957 there
was a difference of over one million tpy between
production capacity and output. [21] Since the
total NH_3 capacity in 1957 was approximately 4.5
million tons, the ammonia industry thus was
operating at only 75 percent of capacity. A 1959

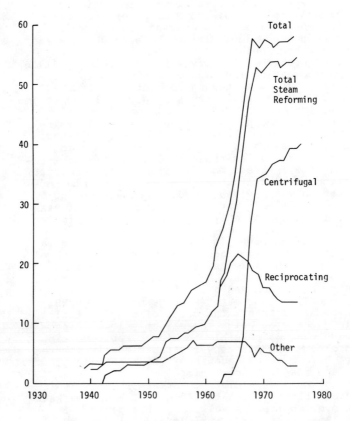

FIGURE 6.2. U.S. Ammonia Production Capacity.
Source: [18].

Chemical Week article refers to the "current general overabundance of ammonia facilities." [22]
As would be expected under such circumstances, few new ammonia complexes were being constructed. Table 6.8 lists the new steam-reforming plants that came on stream in the latter half of the 1950s and the early 1960s. Considering the over-capacity conditions, the lack of an economical means of long distance transportation, and the marginal benefits of a large scale reciprocator-based facility, it is not at all surprising that huge plants were not being built during the late 1950s.
1960, however, was a turning point for a number of these factors. That year saw the growth of refrigerated ammonia storage and transportation at atmospheric pressure; [23] such techniques offer a low cost means of handling large quantities of ammonia. [24] In 1960 two high capacity refrigerated barges for ammonia transport were contracted to be built. [26] The first pipeline for distribution of ammonia was completed in 1962, and a second in 1971. Thus, transportation costs were falling. Moreover, the capacity/output imbalance was closing in the early 1960s. By that time ammonia plant designers felt that markets would soon be able to support larger plants.

Input Coefficients for Steam Reforming

Table 6.9 summarizes the main changes in the nature and conditions of the major unit operations in ammonia plants. Subject to the qualifications on data quality noted in Section 6.1 and elsewhere in this chapter, Table 6.10 presents the changes in the main input requirements of the ammonia steam reforming production process from 1947 to 1972. Elements of Table 6.10 represent the factor input coefficients associated with the most efficient plant size and design available in each year. The published data from which the coefficient time series is derived and an explanation of the methods used to construct the time series can be found in Appendix A, Section A.4.
Conversion to fixed dollars for investment data is based on the Chemical and Engineering News index of plant and equipment with 1957-59 taken as the base years. We assumed an exponent, b, equal to 0.6 to compute a_k, the capital intensity

TABLE 6.8
New Steam-Reforming Ammonia Plants, 1955-1961

Year	Capacity* (tpd)	Owner	Location
1955	200	Escambia Chemical Corp.	Pensacola, Florida
	185	U.S. Industrial Chemicals.Co.	Tuscola, Illinois
1956	114	Northern Chemical Industries	Searsport, Maine
	300	Solar Nitrogen Chemicals Inc.	Lima, Ohio
1957	210	Kaiser Aluminum Corp.	Savannah, Georgia
	200	Phillips Pacific Chemical Co.	Finley, Washington
	200	St. Paul Ammonia Products, Inc.	Pine Bend, Minnesota
1958	--	No New Facilities	
1959	215	E. I. duPont de Nemours & Co.	Gibbstown, New Jersey
	200	Mississippi Chemical Corp.	Pascagoula, Mississippi
	115	Occidental Petroleum Corp.	Lathrop, California
	140	Valley Nitrogen Producers, Inc.	Helm, California
1960	--	No New Facilities	
1961	340	Tennessee Corp.	Tampa, Florida
	200	Farmland Industries, Inc.	Hastings, Nebraska
	215	W.R. Grace & Co.	Woodstock, Tennessee
	300	Solar Nitrogen Chemicals Inc.	Joplin, Missouri
	300	Chevron Chemical Co.	Fort Madison, Iowa

*Assumes 350 production days per year.
Source: [25]

144

TABLE 6.9
Steam-Reforming Operating Characteristics

Year	1946	1957	1959	1961	1963	1965	1967	1969	1971	1972
Pressure to Primary Reformer (psi)	40	200	200	300	300+	400+	400+	350-450	400-500	400-500
Temperature (°F) Primary-Secondary Reformer	1300-1800 ≈1000	1400-1800 700-850	1400-1800 700-850	1400-1800 700-850	1400-1800 700-850	1450-1800 700-850	1450-1800 700-850	1350-1850 650-750	1400-1800 ≈800	1500-1800 ≈800
Shift Reaction					450-550	400-500	400-500	400-475	≈500	≈500
Process Steps CO shift: number of stages/catalyst	1/HT	1/HT	1/HT	1/HT	2/HT	1/HT-LT	1/HT-LT	1/HT-LT	1/HT-LT	1/HT-LT
Compression stage within gas preparation/purification section	yes	yes	yes	no	no	no	no	no	no	no
CO₂ Removal	water	MEA	MEA	MEA	MEA	MEA	MEA	MEA/ Carbonate	MEA/ Carbonate	MEA/ Carbonate
Final Purification System	Copper Liquor	Copper Liquor	Copper Liquor	Copper Liquor	Methanation	Methanation	Methanation	Methanation	Methanation	Methanation
Synthesis Pressure (psi)	5000	4700	4700	4700	4700	4000-5000 or 2000-3000	4000-5000 or 2000-3000	as low as 2000, as high as 9000	2000-4500	2000-4500
Comments		a		b		c			d	

a. Pressure reforming

b. Methanation system of final purification available.

c. Centrifugal compression systems available

d. Features efficient integrated generation of steam, turbine-driven centrifugal compressors, and a high level of process heat recovery.

Source: [28]

145

TABLE 6.10
Data for Regression Analyses*

Year	a_k	a_g	a_e	a_1	X	P_k	P_g	P_e	P_a	W	G	r
1941						3.92	16.7	141	109	74	150	2.84
1942						4.08	17.1	141	109	85	150	2.85
1943						4.13	17.3	143	109	92	1260	2.80
1944						4.12	17.1	144	91	95	1827	2.78
1945						4.10	16.5	143	72	99	1995	2.61
1946						4.43	16.8	144	72	108	2545	2.51
1947						4.93	17.8	145	80	122	2545	2.58
1948	55.90	38.0	1900.0	2.00	150	5.46	18.2	150	94	134	2612	2.80
1949	55.90	38.0	1900.0	2.00	150	5.43	18.3	155	91	142	2762	2.65
1950	55.90	38.0	1900.0	2.00	150	5.62	19.0	151	97	150	2892	2.59
1951	50.87	38.0	1500.0	2.16	200	6.27	20.0	151	97	162	3372	2.84
1952	50.87	33.5	1500.0	2.16	200	6.44	21.0	152	79	169	3659	2.95
1953	37.63	44.9	1416.0	2.50	200	6.77	23.0	158	85	181	5048	3.18
1954	26.77	45.2	858.0	1.47	200	6.95	24.0	158	85	189	6747	2.87
1955	27.72	45.2	676.0	1.47	200	7.08	26.0	158	84	197	7247	3.04
1956	27.71	45.2	676.0	1.30	300	7.88	27.0	160	88	209	7540	3.38
1957	27.71	45.2	676.0	1.30	300	8.82	28.0	162	88	220	8184	3.91
1958	27.71	45.2	676.0	1.30	300	8.80	30.0	164	92	229	8369	3.80
1959	27.71	44.6	676.0	1.30	300	9.57	31.0	164	92	240	9271	4.38
1960	26.70	44.6	658.0	1.11	300	9.60	33.0	165	92	250	9952	4.41
1961	26.70	44.1	658.0	1.11	400	9.55	35.0	167	92	257	11817	4.36
1962	25.93	44.1	642.0	1.11	400	9.48	35.0	168	92	264	13300	4.29
1963	25.93	44.1	642.0	1.11	400	9.42	35.0	172	92	272	16190	4.24
1964	25.11	43.3	618.0	0.95	400	9.65	35.0	171	92	280	16549	4.37
1965	18.73	35.0	50.0	0.18	600	9.85	35.0	171	92	286	22162	4.47
1966	18.23	32.0	30.0	0.12	600	10.83	35.0	170	92	298	29157	5.13
1967	18.27	31.5	30.0	0.08	1000	11.58	35.0	171	92	310	41097	5.53
1968	25.72	31.5	20.0	0.08	1500	12.60	36.0	171	92	326	50557	6.05
1969	25.72	31.5	20.0	0.08	1500	14.20	38.0	172	50	347	57657	6.93
1970	26.34	31.5	20.0	0.08	1500	16.18	41.0	175	59	371	58657	7.84
1971	26.34	30.4	15.2	0.08	1500	16.34	45.0	189	59	393	60157	7.38
1972		30.4	15.2	0.08	1500	16.80		207	65	408	61157	7.26

*Symbols are defined in Table 6.11

146

coefficient. Guthrie, in 1970, reported an exponent
of 0.58 for the ammonia industry. [27]

6.3 EMPIRICAL TESTS OF THE PROCESS
    INNOVATION MODEL

    We report in this section on an attempt to
explain movements in process coefficients and
capacity (Table 6.10) for the steam reforming process over the period 1947-1972. Data sources may
be found in the Appendix, and Table 6.11 defines
symbols used in this section.
    The first subsection states the main hypo-
theses to be tested and the general approach taken.
Results and conclusions from the empirical study
are contained in the remainder of the Chapter.

Hypotheses and Models

    The basic assumption for our empirical analy-
sis is that innovation is, at least partly, responsive to economic considerations; that is,
business firms have some control over the extent
and direction of those changes in production processes which require research and development
effort. In particular, we are concerned with two
aspects of such process changes: those reflected
in input coefficient changes and those reflected in
capacity changes.

    Process Coefficient Equations. With respect
to input coefficients we delineated two types of
innovation in Chapter 2 -- reductions in inputs and
increased substitution possibilities. We concen-
trate on two primary hypotheses for our empirical
tests: 1) input prices will be a determinant of
process coefficients; and 2) improvements in process technology will be greater as more capital
goods are produced (learning by doing) and as
industry growth is expected.
    Accordingly, the models we estimate contain
price variables and cumulative gross capacity in
steam reforming plant and equipment. The latter is
entered in reciprocal form to capture a decreasing
effect from new investment. The capacity variable
should also reflect anticipated industry growth.
We would expect that the own-price regression
coefficients, i.e., the coefficients of P_k, P_g and
P_e in the a_k, a_g, and a_e equations, will be
negative. That is, as the prices of these inputs

147

TABLE 6.11
Definitions of Symbols

Symbol	Definition
q	Chemical Engineering new plant construction index
K	reported plant and equipment costs divided by q
X	plant capacity (tpd)
a_k	capital intensity coefficient = $K/X^{.6}$
a_g	natural gas per ton of ammonia (million BTU)
a_e	electricity per ton of ammonia (KWH)
a_l	labor per ton of ammonia (man-hours)
P_k	price of capital = $q(.05 + $ bond yield$)$ x 100
P_g	price of natural gas per million BTU
P_e	price of electricity per KWH
P_a	price per ton of ammonia
G	cumulative gross capacity (tpd) of steam-reforming plants
CENT	dummy variable for centrifugal compressors. CENT=0 from 1947-64 and 1 from 1965-72
W	wage rate of chemical workers (¢/hour)
REF	dummy variable for introduction of refrigerated storage and transportation: REF=0 from 1947-59 and 1 from 1960-72
$Z(-n)$	indicates an n-year lag in variable Z

are increased, the amount of the input used to produce a ton of ammonia is expected to decrease. It
is difficult to generalize about the signs of the
remaining price coefficients -- the "cross-price"
coefficients such as P_e in the a_g equation. These
will depend on the nature of the relevant technology. Thus, an increase in the price of natural
gas is likely to result in a reduction in a_g;
whether this can be accomplished by reducing a_g
with no effects on other inputs or whether an
increase in one or both of a_k and a_e is necessary
cannot be determined a priori. It depends in a
complex way on the physical and chemical characteristics of the process.

Since it represents both learning by doing and
expected growth, it is anticipated that $(1/G)$ will
enter with a positive coefficient; that is, as
more plants are built and $(1/G)$ decreases, the
amount of each input required per ton is expected
to decrease.

In addition to the price and investment
variables, we included a variable, CENT, which
shifts the equation for the period after the introduction of centrifugal compressors. Because of the
rather large change in a_e which accompanied the
shift from reciprocating to centrifugal compressors,
the newer technology is in some respects more like
a process change than a substitution. We introduced this variable to examine the extent to which
the coefficients of the remaining variables are
affected. It would strengthen the case for the
importance of the price variables if they are
statistically significant when CENT is included.

Capacity Equations. As hypothesized in
Chapter 2, we expect maximum plant capacity to be
limited by technological and market factors.
Accordingly, we examined regressions of maximum
plant size on cumulative capacity, G, to represent
cumulative experience, and on the price of ammonia,
P_a. The latter is intended to give an indication
of the margin for distribution costs when production
is concentrated at one plant to achieve economies
of scale. We did not find it possible to obtain
good data for transportation costs and so were
forced to use a dummy variable, REF, to represent
the period since refrigerated storage and transportation was introduced. Unfortunately, because
we used a six year lag this variable is identical
with CENT, and the two effects are thus confounded.

The capacity equation also included a labor price variable (W) and the price of capital (P_k). Since labor is an increasing-returns factor in process industries, an increase in scale will tend to reduce unit labor costs. Thus, wage increases should be associated with increased capacity. The effect of P_k is more difficult to predict. Increasing capacity will lower unit capital costs, suggesting that increases in P_k lead to increases in capacity. However, increases in P_k will increase total capital costs, leading to substitution against plant and equipment.

Functional Form. In the absence of information regarding the appropriate functional form we experimented with linear and log-linear formulations for the process and capacity equations. The former indicates a constant absolute effect, while the latter implies a constant percentage effect from a change in an independent variable.

In preliminary work we also experimented with ratios of input coefficients and prices. We believe, however, that the ratios tend to mask the separate effects of overall reductions in inputs and substitution between inputs. Accordingly, we present results only for levels and logarithms of levels for each coefficient.

Another aspect of functional form which must be considered is the existence of a time lag between a change in an independent variable and the change in the dependent variable. Such a lag may be present for several reasons: a delay in recognizing price trends which would stimulate substitution or innovation, a lag arising from the time it takes to construct a new plant, or the time period during which R&D is being conducted. In general, a relatively long lag would be evidence that efforts more akin to innovation than to substitution are taking place. Because our time series are short and lagged values of the independent variables are rather highly correlated, we chose to work with simple lag structures. In particular, we estimated models in which each independent variable is lagged the same number of years, and just one lag appears in each model. Since our most clear cut expectations are those regarding the own-price coefficients, we examined how these varied as a function of the time lag. Table 6.12 presents estimates of the own-price coefficients in the most complete models we examined. The model specifications follow:

TABLE 6.12
Lagged Own-Price Coefficients[a]

Lag	α_2	β_3	γ_4
0	-.200 (.278)	-.007 (.004)	1.570 (.603)
1	-.265 (.284)	-.162 (.089)	1.440 (.784)
2	-.106 (.282)	-.198 (.197)	1.556 (.943)
3	.245 (.285)	-.167 (.227)	1.935 (1.292)
4	-.086 (.348)	-.537* (.207)	1.818 (1.704)
5	-.685 (.393)	-.777* (.193)	-2.433 (2.823)
6	-.999* (.409)	-.467* (.184)	-7.269* (3.123)

[a]See text for equations and parameter definitions. Standard errors in parentheses.

*Significant at 5% level.

$$a_g = \alpha_1 + \alpha_2 \, P_g(-n) + \alpha_3 P_e(-n)$$

$$+ \, \alpha_4 P_k(-n) + \alpha_5 \frac{1}{G}(-n) + \alpha_6 \, \text{CENT}$$

$$a_e = \beta_1 + \beta_2 \, P_g(-n) + \beta_3 P_e(-n)$$

$$+ \, \beta_4 P_k(-n) + \beta_5 \frac{1}{G}(-n) + \beta_6 \, \text{CENT}$$

$$a_k = \gamma_1 + \gamma_2 \, P_g(-n) + \gamma_3 P_e(-n)$$

$$+ \, \gamma_4 P_k(-n) + \gamma_5 \frac{1}{G}(-n) + \gamma_6 \, \text{CENT}$$

We computed estimates for n = 1, 2, 3, 4, 5, and 6. Although a somewhat different pattern emerges for each input, note that P_g and P_k do not become negative and significant until n = 6, and each coefficient takes on positive values for occasional earlier lags. P_e is negative and significant starting with n = 4 and remains that way at n = 6. Accordingly, we chose to use n = 6 for the remainder of the empirical work, and these equations are presented below. This result lends support to the view that we are facing an induced innovation phenomenon, rather than a straightforward substitution of one known technology for another.

Because our data base is a time series, we ran the Durbin-Watson test for first-order autocorrelation in the disturbance term. [29] Our tables include the ordinary least squares estimates (OLS) and the generalized least squares estimates (GLS) which correct for the presence of first-order autocorrelation.* It should be noted that the statistical properties of GLS are not established for such small samples as are used in this study.

---

*Calculations were run with the SAS computer program (SAS Institute, Inc., Raleigh, North Carolina).

Regulation and Alternative Hypotheses. The
models discussed in this section do not explicitly
introduce government regulation because the legal
and engineering analyses of earlier chapters and of
Section 6.2 failed to reveal significant technical
developments which may be attributed to regulation.
This is not to assert that there have been no such
impacts, but that it is not feasible to determine
the effects quantitatively for this industry at
this time. Section 7.1 presents our interpretation
of the available evidence on the impacts of regulation on process innovation in ammonia and more
generally.

Two further points regarding the quantitative
analysis of regulation should be briefly noted:
1) the extent to which regulatory actions are
reflected in input prices, and 2) the appropriate
treatment of regulatory changes had such changes
occurred in the time period studied.

Two examples of effects on input prices due
to regulation might be cited. First, certain
environmental and safety regulations may have
caused increases in the prices of capital goods
used in the chemical industry. Such increases will
be reflected in the price index and will affect
innovation in the same way as would any other
price increase. Second, for part of the period
under discussion, government regulation of natural
gas prices may have kept them lower than the prices
which would have prevailed in a free market. (See
Section A.2 of the Appendix for more details.)

To the extent that such prices affect
innovation -- regression results reported below
suggest a significant impact -- the technology
adapted to the price signals. Thus, a possible
indirect effect of natural gas price regulation was
a greater use of natural gas in ammonia production.
However, it should be pointed out that most of the
new capacity in recent years was located along the
Gulf Coast, where unregulated intrastate gas has
been available.

The appropriate econometric treatment of
regulatory change is next briefly discussed within
the framework of our production-innovation model.
Since a change to a new engineering process is
relatively rare, generally well documented, and
widely discussed, it should be relatively straightforward to identify the role, if any, of regulation
in causing the change. Presumably, input coefficient data would be available for both old and new
processes, so that costs could be compared in an

153

attempt to isolate the role of the regulation in bringing about the new process. The detection of the less dramatic types of process change might be investigated by attempting to determine whether a significant change in the regression equation occurred. Statistically, this could be done by fitting separate regressions for each subperiod if sufficient observations are available and by employing standard statistical tests. If not, one might test for the significance of one or more dummy variables which identify periods of different regulatory environments. These two approaches could be employed in equations for both input co-efficients and capacity.

The principal alternative hypotheses which the statistical work is designed to reject is that innovations appear randomly, unrelated to economic forces. It may be that advances in basic science appear in a random fashion, but such advances are not in general immediately useful to industry. A selection of which aspects of scientific knowledge are to be developed must be made and it is our view that this selection is conditioned by profitability prospects.

Another alternative hypothesis is not easily rejected by the data -- the belief that substitution, rather than innovation is being observed. The statistical evidence taken in conjunction with the extended discussion of technical development which appears in Section 6.2 supports the innovation view. This issue is further discussed in Section 7.2.

Regression Results

Process Coefficient Equations. Results of regressions to explain a_k, a_g, a_e, and their respective logarithms may be found in Tables 6.13 to 6.18. The strongest, most robust results are associated with the own-price coefficients, i.e., P_k in the a_k equation, P_g in the a_g equation, and P_e in the a_e equation, and the logarithm forms of each. The estimated coefficients are negative and significant in equations explaining a_g. Negative coefficients are estimated for own-price coefficients in the a_k equations, and all but those estimated in the linear form using GLS were significant. The a_e equations were less clear cut, but are usually negative and occasionally significant. The evidence thus is quite clear that an increase in an input price leads six years later to a

154

TABLE 6.13
Process Coefficient Equations (Standard Errors in Parentheses)

$a_k = \beta_1 + \beta_2 P_k(-6) + \beta_3 P_g(-6) + \beta_4 P_e(-6) + \beta_5 G^{-1}(-6) + \beta_6$ CENT

Estimation Technique[+]	β_1	β_2	β_3	β_4	β_5	β_6	R^2	Degrees of Freedom	F-Statistic	Durbin-Watson Statistic
OLS	169.171 (70.165)	-8.210* (3.163)	2.640* (.801)	-0.933 (.564)			.729	22	19.73*	.739
GLS	115.695 (59.601)	-5.021 (2.646)	1.048 (.835)	-0.475 (.478)			.538	22	8.54*	
OLS	110.922 (71.577)	-7.304* (2.991)	1.965* (.819)	-0.502 (.568)	1.840 (.903)		.774	21	17.95*	.853
GLS	87.231 (63.023)	-4.835 (2.633)	.892 (.837)	-0.285 (.498)	1.287 (.838)		.603	21	7.97*	
OLS	110.527 (73.637)	-7.269* (3.123)	1.927 (1.055)	-0.496 (.592)	1.840 (.925)	.348 (5.834)	.774	20	13.68*	.850
GLS	86.194 (64.714)	-4.874 (2.721)	.841 (.911)	-0.270 (.518)	1.273 (.860)	.767 5.518	.601	20	6.04	

[+]OLS is ordinary least squares; GLS indicates that the estimates have been corrected for first-order autocorrelation by generalized least squares.

*Significant at 5% level.

155

TABLE 6.14
Process Coefficient Equations (Standard Errors in Parentheses)

$a_g = \beta_1 + \beta_2 P_k(-6) + \beta_3 P_g(-6) + \beta_4 P_e(-6) + \beta_5 \frac{1}{5G}(-6) + \beta_6 CENT$

Estimation Technique[+]	β_1	β_2	β_3	β_4	β_5	β_6	R^2	Degrees of Freedom	F-Statistic	Durbin-Watson Statistic
OLS	-18.988 (39.209)	-3.679* (1.767)	-2.357* (.447)	.582 (.315)			.660	22	14.26*	1.22
GLS	-9.786 (39.757)	2.272 (1.775)	-1.832* (.506)	.501 (.320)			.502	22	7.39*	
OLS	-18.556 (43.773)	3.672 (1.829)	-2.352* (.501)	.578 (.347)	-.014 (.552)		.660	21	10.21*	1.22
GLS	-3.680 (43.887)	2.120 (1.826)	-1.766* (.545)	.457 (.348)	-.204 (.566)		.504	21	5.34*	
OLS	-4.535 (28.500)	2.407 (1.211)	-.999* (.409)	.354 (.229)	-.023 (.359)	-12.342* (2.262)	.863	20	25.30*	1.61
GLS	1.052 (29.625)	2.334 (1.237)	-.927* (.419)	.310 (.238)	-.058 (.374)	-12.291* (2.361)	.834	20	20.02*	

[+]OLS is ordinary least squares; GLS indicates that the estimates have been corrected for first-order autocorrelation by generalized least squares.

*Significant at 5% level.

TABLE 6.15
Process Coefficient Equations (Standard Errors in Parentheses)

$a_e \times 100 = \beta_1 + \beta_2 P_k(-6) + \beta_3 P_g(-6) + \beta_4 P_e(-6) + \beta_{5G}^{1}(-6) + \beta_6$ CENT

Estimation Technique[+]	β_1	β_2	β_3	β_4	β_5	β_6	R^2	Degrees of Freedom	F-Statistic	Durbin-Watson Statistic
OLS	89.730* (25.211)	-2.049 (1.136)	.609* (.288)	-.533* (.203)			.875	22	51.37*	.841*
GLS	55.264* (22.205)	-2.076* (.985)	.223 (.313)	-.248 (.178)			.753	22	22.34*	
OLS	69.148* (25.802)	-1.729 (1.078)	.370 (.295)	-.381 (.205)	.650 (.325)		.895	21	44.76	1.007
GLS	48.276 (23.804)	-1.967 (.994)	.183 (.315)	-.205 (.188)	.423 (.316)		.796	21	20.52*	
OLS	74.521* (22.949)	-2.213* (.973)	.889* (.329)	-.467* (.184)	.647* (.288)	-4.729* (1.818)	.922	20	46.99*	1.318
GLS	62.383* (23.220)	-1.931 (.960)	.608 (.324)	-.358 (.186)	.568 (.297)	-3.938 (1.888)	.881	20	29.49*	

[+]OLS is ordinary least squares; GLS indicates that the estimates have been corrected for first-order autocorrelation by generalized least squares.

*Significant at 5% level.

TABLE 6.16
Process Coefficient Equations (Standard Errors in Parentheses)

$$\ln a_k = \beta_1 + \beta_2 \ln P_k(-6) + \beta_3 \ln P_g(-6) + \beta_4 \ln P_e(-6) + \beta_5 \ln \frac{1}{G}(-6) + \beta_6 \text{ CENT}$$

Estimation Technique[+]	β_1	β_2	β_3	β_4	β_5	β_6	R^2	Degrees of Freedom	F-Statistic	Durbin-Watson Statistic
OLS	-1.330 (10.839)	-2.110* (.484)	1.336* (.468)	.887 (2.366)			.801	22	29.51*	.838
GLS	.956 (9.614)	-1.551* (.496)	.785 (.560)	.574 (2.109)			.648	22	13.49*	
OLS	-4.631 (12.036)	-2.020* (.508)	1.228* (.501)	1.586 (2.615)	.039 (.059)		.805	21	21.69*	.839
GLS	-1.342 (10.467)	-1.468* (.525)	.732 (.576)	1.041 (2.277)	.036 (.059)		.657	21	10.03*	
OLS	-5.491 (12.230)	-2.163* (.550)	1.493* (.624)	1.649 (2.646)	.037 (.060)	-.098 (.134)	.810	20	17.07*	.884
GLS	-1.480 (10.868)	-1.555* (.536)	.929 (.636)	.981 (2.369)	.037 (.060)	-.064 (.137)	.684	20	8.64*	

[+]OLS is ordinary least squares; GLS indicates that the estimates have been corrected for first-order autocorrelation by generalized least squares.

*Significant at 5% level.

158

TABLE 6.17
Process Coefficient Equations (Standard Errors in Parentheses)

$$\ln a_g = \beta_1 + \beta_2 \ln P_k(-6) + \beta_3 \ln P_g(-6) + \beta_4 \ln P_e(-6) + \beta_5 \ln \tfrac{1}{G}(-6) + \beta_6 \text{ CENT}$$

Estimation Technique[+]	β_1	β_2	β_3	β_4	β_5	β_6	R^2	Degrees of Freedom	F-Statistic	Durbin-Watson Statistic
OLS	6.583 (6.983)	1.051* (.312)	-1.448* (.301)	-.065 (1.524)			.628	22	12.37*	1.012*
GLS	4.228 (6.540)	.662 (.328)	-1.105* (.361)	.331 (1.433)			.411	22	5.12*	
OLS	3.859 (7.701)	1.125* (.325)	-1.538* (.320)	.513 (1.673)	.033 (.038)		.641	21	9.35*	1.094
GLS	3.273 (7.252)	.749* (.348)	-1.187* (.371)	.544 (1.578)	.016 (.039)		.444	21	4.19*	
OLS	.972 (4.255)	.644* (.192)	-.647* (.217)	.723 (.921)	.025 (.021)	-.328* (.047)	.896	20	34.64*	1.965
GLS	.744 (4.178)	.647* (.187)	-.657* (.211)	.744 (.903)	.026 (.020)	-.329* (.045)	.905	20	38.09*	

[+]OLS is ordinary least squares; GLS indicates that the estimates have been corrected for first-order autocorrelation by generalized least squares.

*Significant at 5% level.

TABLE 6.18
Process Coefficient Equations (Standard Errors in Parentheses)

$$\ln a_e = \beta_1 + \beta_2 \ln P_k(-6) + \beta_3 \ln P_g(-6) + \beta_4 \ln P_e(-6) + \beta_5 \ln \tfrac{1}{G}(-6) + \beta_6 \text{ CENT}$$

Estimation Technique[†]	β_1	β_2	β_3	β_4	β_5	β_6	R^2	Degrees of Freedom	F-Statistic	Durbin-Watson Statistic
OLS	91.814 (55.988)	3.346 (2.500)	-5.877* (2.415)	-14.615 (12.224)			.817	22	32.67*	.649*
GLS	17.131 (39.224)	-1.469 (2.171)	-3.868 (2.599)	.705 (8.621)			.643	22	13.20*	
OLS	78.645 (62.439)	3.706 (2.637)	-6.309* (2.596)	-11.824 (13.565)	.157 (.306)		.819	21	23.75*	.663*
GLS	15.400 (42.896)	-1.359 (2.293)	-3.926 (2.666)	1.053 (9.321)	.035 (.258)		.646	21	9.60*	
OLS	51.892* (17.020)	-.751 (.766)	1.943* (.868)	-9.876* (3.683)	.089 (.083)	-3.043* (.187)	.987	20	311.14*	1.674
GLS	46.226* (17.297)	-.692 (.792)	1.487 (.912)	-8.493* (3.752)	.094 (.087)	-2.967 (.196)	.984	20	250.66*	

[†]OLS is ordinary least squares; GLS indicates that the estimates have been corrected for first-order autocorrelation by generalized least squares.

*Significant at 5% level.

decrease in the use of that input. Although the
statistical results are consistent with a hypothesis of input substitution, we argued at length
in earlier sections that the observed input coefficient changes required research and development
efforts. We therefore conclude that our primary
hypothesis stating a negative relationship between
a lagged input price and the associated input coefficient is consistent with the data.

With a few exceptions, the cross-price
coefficients failed to yield clear cut conclusions;
some results were extremely sensitive to model
specification. The relationship between capital
and gas inputs was the most consistent. Gas prices
entered with a positive sign in the a_k equation,
as did P_k in the a_g equation. That is, increases
in gas prices lead to increases in capital intensity, and increases in the price of capital lead
to more gas input. This suggests a technical
relation of substitutability between these two
inputs. This relationship holds up in both linear
and log-linear forms, and is statistically significant in a number of equations. Moreover, the
introduction of CENT does not greatly affect the
results.

The remainder of the cross effects are highly
dependent on functional form. The sign of the
relationship between electricity prices and capital
input reverses when the specification changes from
linear to log-linear, as does the relationship
between capital prices and electricity input.
Similar mixed results are found with gas and
electricity prices and inputs. However, many of
these coefficients are not significantly different
from zero. More work needs to be done in this
area.

Except for the linear a_g equation, cumulative
investment has the expected sign, but is not
statistically significant. That is, an increase in
G leads to reduction in a_k, a_e, and the logarithms
of a_k, a_g, and a_e. However, the coefficients are
not sufficiently large to be distinguished from
zero.

The CENT variable enters negatively and significantly in both forms of the a_e and a_g equations,
and negatively, but insignificantly, in the a_k and
ln a_k equations. The shift to centrifugal compressors, therefore, was associated with a
reduction in all inputs. Nevertheless, the own-
price coefficients remain negative and significant
when CENT is included, indicating that the

161

significant impact of own-prices is not the
spurious result of a coincidental rise in input
prices and the introduction of centrifugal compressors.

Capacity Equations. Regression results for
the capacity equations may be found in Table 6.19.
Although few of the coefficients are statistically
different from zero, the expected signs are obtained: P_a enters positively, indicating that the
size of market for a plant increases with the price
of the final product; G enters positively, reflecting the effect of cumulative experience on
technology; W enters positively since increases in
scale tend to reduce unit labor costs; and REF is
positive and significant. Unfortunately we cannot
unequivocally attribute the latter to the effect of
improved transportation facilities on expanding
the market since the dummy variable representing
REF is equal to CENT, and the introduction of
centrifugal compressors was associated with a large
increase in capacity. Finally, although results
with P_k are not clear cut, it appears that the
effect is generally negative, suggesting that
attempts to substitute against capital equipment
when P_k rises offset attempts to reduce unit costs
by increasing capacity.

6.4 CONCLUSIONS

To our knowledge, the foregoing analysis is
the first attempt to examine engineering process
coefficients for the purpose of explaining technical
progress as a function of input prices and other
variables. Our results, although tentative,
suggest that the effort is worthwhile. With respect to hypotheses concerning the relationship
between an input and its own price, the regression
results were quite favorable. In other cases, the
results were in the expected direction, although
not clear cut. We believe that additional efforts
with these data and similar data for other
industries is warranted.
In any event, the results are consistent with
an endogenous view of process change, although
whether this is interpreted as innovation or substitution cannot be settled from the regression
results alone. The qualitative analysis of Section
6.2 and the fact that the six-year lag proved

TABLE 6.19
Capacity Equations[+] (Standard Errors in Parentheses)

$$X = \beta_1 + \beta_2 \, P_a(-6) + \beta_3 \, G(-6) + \beta_4 \, W(-6) + \beta_5 \, P_k(-6) + \beta_6 \, REF$$

β_1	β_2	β_3	β_4	β_5	β_6	R^2	Degrees of Freedom	F-Statistics	Durbin-Watson Statistics
-115.301 (307.817)	3.926 (2.897)	.016 (.009)	10.221* (3.139)	-259.327* (88.232)	609.802* (109.483)	.947	20	71.14*	1.651
-787.196 (441.428)	7.740 (4.387)	.034* (.014)	9.627 (4.890)	-190.413 (136.172)		.864	21	33.41*	.424*

$$\ln X = \beta_1 + \beta_2 \ln P_a(-6) + \beta_3 \ln G(-6) + \beta_4 \ln W(-6) + \beta_5 \ln P_k(-6) + \beta_6 \, REF$$

β_1	β_2	β_3	β_4	β_5	β_6	R^2	Degrees of Freedom	F-Statistics	Durbin-Watson Statistics
-5.254 (3.034)	.528 (.441)	.004 (.117)	2.172* (.943)	-1.323 (.948)	.848* (.129)	.959	20	94.69*	1.957
-6.821 (5.257)	1.409 (.731)	.207 (.197)	.678 (1.591)	.700 (1.558)		.871	21	35.60*	.571*

[+]Equations are estimated by ordinary least squares.

*Significant at 5% level.

163

superior to shorter lags point in the direction of
an innovation interpretation.

7. Conclusions and Interpretations

We have attempted in this study to determine the impact of changes in input prices and government regulations on the processes used to produce ammonia, a product that is typical of the heavy chemical industry. Our approach has been interdisciplinary, incorporating legal, engineering, and economic analysis.

Legal analysis was used to determine the set of regulations faced by producers over the relevant period of time as the background for possible impacts. In addition, the analysis leads to a number of overall conclusions and interpretations which are presented below.

Engineering analysis was used to describe the various processes, to interpret process change in terms of the nature of the innovation, to quantify input and other process characteristics, and to examine environmental problems and their possible solutions.

Economic analysis was used to develop and test a model of production which appears to be consistent with the view of engineering and which provides a taxonomy for process innovation.

We first offer interpretative comments regarding the impacts on innovation of the environmental and workplace regulations and then turn to input price implications. We close with suggestions for future analysis and for policy.

7.1 IMPACTS OF ENVIRONMENTAL AND
WORKPLACE REGULATIONS ON INNOVATION
IN THE AMMONIA MANUFACTURING
INDUSTRY: AN INTERPRETATION

A review of Chapter 3 suggests that a substantial number of environmental and workplace
health and safety regulations apply to the ammonia
manufacturing industry. However, taking account
of Chapters 3 and 5, we conclude that, at least
until very recently, the standards imposed by
environmental regulations have been consistent
with standards to which the industry has been conforming for other reasons, and the additional
costs of meeting more recent standards has been
small. Further, we did not attempt to describe
or analyze the influence of workplace regulations
on ammonia technology, because by all indications
no important influence has existed. Thus, notwithstanding their presence we conclude that
neither the environmental nor the workplace
regulations have played a decisive role for the
innovations which have occurred in the industry.
Nevertheless, a number of observations can
be made which suggest interesting relationships
between regulation and innovation for the industry.
We next examine six major aspects of these relationships: (1) the importance of the environmental
and workplace regulations for the industry in
recent experience; (2) the importance of environmental and workplace regulations for the industry
in its early years; (3) the historical consis-
tency of the regulations with existing consensus
standards; (4) the relative impact of "design"
and "performance" standard regulations on
industry's efforts to comply; (5) the impact of
EPA's guidelines on technology used to achieve
compliance; and (6) the implications of OSHA's
multi-tiered enforcement mechanism.

The Importance of the Environment and
    Workplace Regulations for Industry:
    The Recent Experience

The Fertilizer Institute and individual firms
devote considerable attention to the environmental
and workplace regulations imposed upon them.
Further, a number of people within OSHA and EPA
have developed substantial expertise with the
industry's problems. Relative to other industries,
however, the nitrogen fertilizer industry has not

166

been characterized by longstanding heavily imposed regulations, nor have its workplace dangers and pollutants been the subject of intense public scrutiny. If anything, it has garnered considerable public and congressional support from its vital contribution to food production. Hence, the regulations we consider are a minor factor in the overall operations of the industry.

That is not to say, however, that the regulations do not represent some impositions on the industry. The regulations have increased costs for the industry and added an extra set of considerations for management. The increased costs involve both capital expenditures, such as for water treatment facilities, and operating expenditures for monitoring and reporting. Historically, early local smoke and particulate regulations probably affected the siting of plants, but not operating or capital costs to any significant extent.

In this context, it seems likely that regulations may have been a consideration for deciding whether to introduce process innovation, but that they were not the controlling factor. Consider the introduction of centrifugal compressors: we know that they were substituted for reciprocating compressors and that centrifugal compressors are less dangerous for the work force. It does not appear, however, that concern for workplace safety -- whether stimulated by regulation or workmen's compensation laws or something else -- was a significant factor in introducing the centrifugal compressors. The literature on the industry does not suggest that special problems in either workplace or environmental areas had grasped the attention of the industry or engineering firms working for it, and that literature does not suggest that any particular thought was given to the environmental or workplace significance of the various innovations. On the other hand, the elimination of the dangerous "push-pull" characteristics of reciprocating compressors could well have been recognized by industry as a secondary benefit of eliminating those compressors.

More recent proposed regulations may be much greater impositions than regulations have been historically. The industry's spokesmen seem to be taking this view with respect to EPA's consideration of a zero discharge standard for nitric acid and OSHA's consideration of a 50-ppm

167

ammonia exposure limit. Whether this is in fact
the case cannot objectively be determined unless
and until those agencies impose the standards and
observations can be made of the industry's response
to them. Obviously, we cannot assess that now.

Ultimately, nitrogen fertilizer use may have
to be controlled or reduced in order to avoid a
possible threat to the stability of the ozone
layer. Would such control be thought necessary,
the shock to the ammonia industry would be great,
but without substitute fertilizers the impact on
American agriculture and eating habits would be
even greater.

The Importance of Environmental and
Workplace Regulations for the
Industry: The Early Years

In the early years of the industry's development, say 1918 to 1940, the potential interests
of regulators were coincident with those of good
business and engineering practice. Because of the
very high pressures employed, the acute toxicity
of such gases as carbon monoxide and ammonia, and
the extreme flammability of hydrogen mixtures, it
was essential that plants be designed with clear
regard for the danger of leaks or accidents --
i.e., for a high degree of safety. No one ignores
a gas leak at 4000 psi. Reflecting the lack of
general concern for chronic toxicity of the times,
it was no doubt assumed that the pungent odor of
ammonia which can be detected at 50 ppm was
sufficient warning to workmen. Further, no
instrument was available for routine measurement
at that level.

Through the early years, the only two large
U.S. plants based on coke were in operation at
Hopewell, Virginia, and Belle, West Virginia.
The air pollution which no doubt existed at
Hopewell was not near a center of population, and
that from Belle was probably considered just
another contributor to the miasma of the Kanawha
Valley.

The shift from coke-based plants to the
modern process of steam reforming of natural gas
in World War II was undoubtedly accompanied by
very significant decreases in air and water
pollution, solids disposal problems, and land
despoilage. But the literature of the time makes
no reference to these environmental improvements
as a driving force for the shift. The comparative

economics of the systems based on coal and on
natural gas give the overwhelming nod to gas, even
at 1977 prices, let alone at 1940 prices when gas
was nearly free. Unfortunately, we have no data
that would allow us to compare these processes on
environmental, health, or safety grounds. However,
the shift to the gas technology was hastened by
the federal government's decision to use the new
technology for the war plants it later sold to
private industry. It is unlikely that a concern
with air pollution was an important factor in
those tumultuous years, but we have not been able
to ascertain why the new technology was adopted
in the middle of the war effort.

The industry is again looking seriously at
coal as a potential feedstock, and it is clear that
the environmental problems of coal handling and
processing are significant barriers to a return
to coal, just as they are for many other large
sized industrial boilers.

The Historical Consistency of Some Regulations with Existing Consensus Standards

Throughout the review of workplace regulations
(the regulations which have applied to the industry
for a long time), we observed that the standards
applied were generally drawn from standards promulgated by various professional code-setting
institutions. For example, standards of the
American National Standards Institute, the American
Society of Mechanical Engineers, and the National
Fire Protection Association have all been applied
to this industry. Traditionally, these institutions bring together top professionals who design
and operate the equipment which is the subject of
the standards involved, and they often also involve
such state and local officials as fire marshals
and fire chiefs. As a result, the codes developed
by these institutions reflect a consensus of
appropriate safety standards within economic constraints. In such circumstances, new process innovations are often developed by professionals who
subscribe to the institutional standards as a
matter of good professional practice. This implies
that it would often be an exercise in sophism to
attempt to distinguish decisions made by designers
and operators based upon code standards from those
based upon good professional practice.

169

Some innovations will represent a departure from generally accepted engineering or other professional standards. In such circumstances industry codes may stall introduction of an innovation. Even then, it may be difficult to distinguish the regulation based on such a code from commercial forces as the cause for inhibiting the introduction of the innovation. Codes are used by insurance companies to set standards for safe operation as a condition to underwriting, and compliance with code standards is often an index of a person's professionalism. Thus, a firm will be constrained from departing from code regardless of the regulation's adoption of the code.

All this is not to say that the adoption of codes as standards by OSHA and other regulatory agencies is an unnecessary exercise. Doing so provides an additional enforcement device to ensure that all parts of the industry are complying with generally agreed upon standards. The point is that in such situations it would be erroneous to perceive that the regulations also are constraining innovation. If the standards are constraining useful innovation and if the standards are based on professional practice, it is more accurate to perceive the constraint as one derived from the industry's professional standards.

In one circumstance such regulations may constrain innovation. If the regulations adopt a specific code and the institutional author of that code subsequently adopts a different code, then the old code would continue to have force only because of the regulation. If the old code constrains innovation, then the regulation is constraining innovation. We have seen an example of this occurring in ammonia production. It will be recalled that the Fertilizer Institute had contended for modification of the ammonia storage and handling regulation largely on the grounds that the existing regulation is inconsistent with current industry code standards. (See Section 3.3.) Retaining the old code would be appropriate if the agency has made a conscious determination that the new code is inadequate. Otherwise the only explanation for the situation is slowness of agency responsiveness, and that would constitute an unwarranted regulation imposition.

170

The Relative Impact of "Design" and "Performance" Standards Regulations on Industry's Efforts to Comply

Conventional wisdom suggests that regulatory standards can be categorized into those setting performance standards and those setting design standards. It is often contended that performance characteristic regulation is preferable since it will produce the desired result with fewer restrictions upon the manner in which a firm complies. Our analysis suggests that the distinction between these categories is deceptive and that the terms have different meanings for economists than they have for engineers and lawyers.

Speaking abstractly, design standard regulations are those which specify the design of a piece of equipment or process. For example, it will be recalled that OSHA requires anhydrous ammonia containers to have a "shell or head thickness . . . not less than three sixteenths inch." (See Section 3.3.) This is to be compared with performance standard regulations, that is, those which describe the performance characteristics of a process or piece of equipment and allow any technical solution which has those characteristics. For example, such a regulation might permit any design which performs without danger to workers or the environment under specified conditions of pressure, temperature, chemical exposure, and so forth. One ascertains compliance with design standards by examining whether fabrication was according to design; one ascertains compliance with performance standards by testing the actual process or equipment under real or simulated conditions.

The difficulty with this abstract distinction lies in the fact that such abstract performance standards are ordinarily extremely costly to implement in practice. The process or equipment must be developed virtually to completion before it can be ascertained whether it complies with the performance standards. As a result, the regulations often preferred by engineers and lawyers differ from "pure" performance standard regulations. They prefer regulations based on laboratory tests and calculations that must be passed by the candidate process or equipment and by its construction materials or other constitutents. For example, the tests might include minimum tensile, shear, and impact strengths to be achieved by specimens of the proposed construction materials under conditions of temperature and chemical environment to be expected in practice.

In one sense these "modified" performance standards are merely a variation on design standards since they state, in slightly more abstract terms than the traditional design standard regulation, the design characteristics required for the new process or equipment. They are superior to design standard regulations from the regulated firm's perspective since they allow more flexibility for complying with regulations, and they are superior to the traditional performance standard regulation since they allow greater certainty about the permissibility of using the innovative process or equipment before full development investments have been made.

On the other hand, an agency may rightfully resist these modified performance regulations if there is an insufficient basis for concluding that the standards will prevent introduction of those innovations which are inconsistent with the environmental or workplace health and safety goals. This possibility is considerably more likely in the case of modified performance regulations than in the case of those based on either design or pure performance standards. The success of the modified performance regulations depends upon whether the tests employed and/or calculations made will successfully differentiate innovations which will violate regulatory goals from others. Basic to the engineering discipline is the postulate that tests and calculations made without experience in the actual use of a process or piece of equipment cannot accurately determine how the process or equipment will operate in real or even simulated conditions. This problem is eliminated, by definition, for pure performance standards. It is obviated for design standards since the design standard is a description of a process or piece of equipment with which there has been actual experience.

As a result, we conclude that modified performance standards present fewer barriers to innovation than design or pure performance standards, but we also conclude that modified performance standards introduce a greater risk that the standards will not produce the result for which the regulation has been imposed. This suggests that the pattern we observed in the OSHA storage and handling regulation was judicious.

First, design standards were imposed; then efforts
were undertaken to develop modified performance
standards for which there was a good likelihood
the standards adopted would produce the desired
regulatory goal -- in this case, containers which
are not likely to leak ammonia at the workplace.

The Impact of EPA's Guidelines on
Technology Used to Achieve Compliance

As a general proposition, ammonia plants were
not constructed with a view to holding effluent to
near zero discharge levels. As a result, EPA
administrators are convinced that there are substantial possibilities for adjusting the manufacturing process in order to reduce effluent and
to do so more cheaply than by adding "end-of-pipe"
clean-up equipment at the plant's outflow. If
regulations are designed to allow clean-up by
process modification, then firms will be able to
comply with less expense. Concommitantly, EPA
administrators observe, higher standards can be
imposed upon a facility which complies by in-
process adjustments than upon one which adopts
an end-of-pipe solution, assuming the expenses
involved in compliance in both cases are the same.
This situation leads EPA to attempt to use
performance rather than design standards for
regulations. This is particularly necessitated by
the fact that the in-process adjustments will
differ for each plant because of variations in
design. According to EPA administrators, industry
representatives have resisted such standards
because they do not help industry understand how
to comply and they may impose an added burden on
industry. For example, compliance with a standard
which requires that a firm remove ammonia from
effluent by the use of a specific end-of-pipe
device, a steam stripper, say, is achieved by
installing and operating the device. Use of the
device may not reduce effluent levels to the
degree that EPA had hoped when it required the
device, but the firm has complied with the
regulation regardless. If the standard sets a
maximum allowable ammonia concentration in the
effluent, compliance is achieved only by meeting
the effluent standard, regardless of what devices
or techniques are used.
Industry's position has been that EPA should
include descriptions of how compliance might
reasonably be achieved for any standard which is

stated in performance terms. Since end-of-pipe devices can be described generically, but process adjustments cannot, a requirement that EPA explain how to comply with a standard has the impact of restricting standards to those which can be achieved with end-of-pipe solutions at a cost which is consistent with that anticipated by the legislation. For those firms which use process adjustments to meet the standards, the expense of compliance is reduced.

One possible solution for EPA is promulgating performance standards and publishing examples of ways that those standards have been met at some plants as a guide for process adjustments for noncomplying plants. Such a solution may not be acceptable because EPA cannot reach firm conclusions about the cost of meeting such standards at each plant. If EPA is blocked from imposing such standards, however, it will be required to continue emphasis on the more expensive end-of-pipe solutions. That in turn would weaken a technology-forcing aspect of the pollution control regulations, thus retarding the agency's ability to force innovation in the direction of process adjustment.

The Implications of OSHA's Multi-Tiered Enforcement Mechanism

In Section 3.3 we discussed a rather complex enforcement program for OSHA standards. We concluded that the net effect of the program is to put the heaviest emphasis on encouraging compliance with those standards which are generally accepted as needed for workplace health and safety and those standards which involve matters likely to be considered very significant by the workforce. The remaining question is what impact that sort of standard has for innovation in the workplace.

Generalization about this particular question is extremely dangerous. One of the features of the multi-tiered enforcement mechanism is that strict enforcement is imposed on some companies and none on others. Further, the strict enforcement flows from the presence of an OSHA inspector at a plant, and the inspectors appear more or less randomly. Moreover, it should be assumed that large numbers of companies subject to regulations will comply with those regulations with great rigor as a matter of good citizenship and regardless of the threat of enforcement.

174

Notwithstanding, it does seem that this
complex enforcement mechanism does encourage compliance with important standards and leaves some
room for independent judgment by firms with respect to standards. That is, the firms have some
leeway to interpret vaguenesses in the regulations
and to make their own judgment of what constitutes
compliance. Under the circumstances this suggests
that firms may be less inhibited from changing
production processes and introducing innovations
than they would if they faced fines with near
certainty every time that they vary from OSHA's
interpretation of what the regulations require.

7.2 IMPLICATIONS REGARDING INPUT
     PRICES AND INNOVATION

     We have been able to establish that statis-
tically significant relationships exist between
input factor prices and the amounts of those
factors used in the newest ammonia plants. The
existence of six-year lags between factor price
changes and factor inputs suggests strongly that
more than simple substitution is involved; the
delay is caused by some combination of the need to
do research, the need to wait until further expansion is warranted, and the lag between market
signals and the opening of a new plant.
     Strongest relationships on the whole were
observed between the use of a factor and its own
price; that is, innovation appears to have been
biased in the direction of replacing the newly
expensive factor, rather than on reducing production costs generally.
     No clear way emerged by which the available
information on environmental and workplace regulation experiences could be integrated into the
empirical model, other than directly through
factor price changes. This result is due in large
part to the fact that the technology employed in
the production of ammonia to date has not been
heavily influenced by regulation, as discussed
in Chapters 3 and 5. Modeling techniques which
might be employed had regulation proved more
important are discussed below.
     Innovative activity toward new ammonia pro-
cesses occurs in a complex set of institutions.
This report, and Helscher's thesis, identify many
contributors to ammonia innovation including large
chemical companies, small producers and

cooperatives, equipment suppliers, catalyst
suppliers, design and engineering firms, foreign
firms, the federal government, and producers of
related hydrogen-based products. In view of the
many actors in ammonia innovation, and in view of
the absence of a data base on R&D activity for any
or all of these, we did not attempt to quantify
innovative activity for the purpose of relating it
to driving forces or to innovation.

For purposes of this study innovation was
deemed to have taken place when substantial
research and development was undertaken in order
to change a production process. Because we were
not able to acquire data on R&D expenditures, we
inferred such efforts from other sources. For
example, reforming operating pressures at the end
of the period we observed (1972) were substantially
higher than those at the beginning. Moreover,
there is evidence that the higher pressures
economize on all inputs. Thus we conclude that
the shift to higher pressures required innovative
effort. In addition, wherever possible we quoted
industry sources regarding the extent of innovative
efforts.

To some extent, the dispute over whether
there has been an innovation or a substitution is
semantic: it may be argued that expenditure on
R&D when there is reasonable certainty regarding
the outcome is conceptually similar to the purchase of new equipment; i.e., it is an investment,
which though not a costless means of substitution,
is nonetheless substitution. Perhaps the distinction turns on the extent of uncertainty
regarding the outcome of R&D efforts. We have
argued for the position that the changes in the
steam reforming process required substantial
research efforts in such fields as metallurgy and
chemistry; they involved efforts to improve unit
operations on the one hand and overall system
operation on the other. The latter required
sophisticated computer modeling to establish
tradeoffs between unit efficiency and system
optimization. Finally, catalyst improvements
contributed greatly to process development, and
such improvements require analytic as well as
trial-and-error techniques.

We did not develop a model in which innovations
become available under uncertainty. However, the
standard regression model we employed provides for
random error. The significance of the own-price
coefficients indicates that our results are

176

consistent with the view of innovative efforts
which are responsive to such economic signals as
input price movements. As the data series are
extended it may be possible to look for such response across processes.

Mildly encouraging results were found for our
hypothesis concerning "learning by doing" in capital goods production. More investigation should
be undertaken with this concept in order to
determine the most satisfactory proxy variables
for regression models.

The assumption of increasing returns to scale
for certain inputs -- notably plants, equipment,
and labor -- is consistent with the observed growth
in plant capacity in ammonia, as well as other
chemical products. Our attempts to model the
growth of capacity in steam-reforming plants met
with mixed success. The strongest effects came
from the wage rate variable; ammonia price exerted
a slight influence. A good data series for a
very promising variable, transportation costs, is
not available at this writing.

7.3 SUGGESTIONS FOR FURTHER RESEARCH

Two principal questions arise in connection
with further research in this area: first, how
can the study of innovation in ammonia processes
be extended or improved, and, second, what additional work on process innovation might prove
fruitful?

Several paths might be followed to improve the
present study. Although we searched the open
literature diligently, it is reasonable to believe
that older factor input data may be available from
industry or government files. It would be
desirable to develop enough additional data points
prior to World War II to attempt to push the
regression model back to cover the period of
transition from coal to gas as feedstock. Data
for this purpose may be contained in the files of
the War Production Board at the National Archives.

Likewise, since a number of new plants have
recently been announced, including several based
on coal feedstock, it would be interesting to
acquire sufficient data to complete the data base
from 1972 to the present. One test of our
empirical model would be to use it to predict the
characteristics of the new technology based on the
estimated coefficients.

177

The largest unresolved part of the present study is to include transportation costs and local ammonia market sizes as constraints on maximum plant size and therefore on technology selection. This might be accomplished in a modified recursive programming format, the details of which we have not developed. Unlike existing uses of that method the study we suggest would attempt to explain not only the selection of technology, but generation of the alternatives.

We believe it would be of interest to carry out analyses similar to those done in this study for other products for which regulation has been more significant. However, the selection of such a product is not obvious, and it might be assisted by one or two surveys. For example, one might survey a population of producers, regulators, and academics for nominations and rankings of products for which processes have changed and for which process regulation has been significant. A similar survey might be attempted using the method described in Section 5.5 in which a count of citations would be used to identify products for which both innovation and regulation are of major concern. Other criteria must be used, of course, including data availability, a long history, and an unchanging product.

Should such products be identified, regulatory driving forces could be examined in our model directly, by using measures of regulatory activity as explanatory variables (number of violations, enforcement actions, etc.), or indirectly, by using dummy variables to identify pre- and post-regulatory periods. The latter approach has been used in studies of the impacts of regulation on drug innovation.

Information on patents is another possible data source. Patent activity associated with a particular product may be useful both for identifying products which are experiencing pressure on costs from regulations or price changes and for measuring the extent of innovative response to such pressure.

7.4 POLICY-RELATED FINDINGS

In the pursuit of its goals, actions of the
federal government often affect the ammonia
industry and chemical process industries in general.
Our analysis has uncovered several findings which
may be useful in the formulation of future policy
towards industry. These are summarized in this
section.

(1) The federal government has played an
important role in the growth of ammonia
production, the structure of the industry,
and in the development of new technology
for its production.

The ammonia industry was born and rushed
through its early development stages as a result
of government actions in World War I and II.
These actions were taken to assure the availability of nitrogen for fertilizers and explosives.
Subsequently, TVA and the Agricultural Extension
Service stimulated its use through promotional
activity, and TVA continued to do research on
process technology. U.S. agricultural policy has
also stimulated fertilizer use by a combination of
price supports and acreage restrictions. In
recent years, natural gas price regulation has
served to keep gas prices low, making ammonia
attractive as an agricultural input. The
government's sale of its World War II ammonia
plants to producers which had not previously produced ammonia increased competition in the
industry.

(2) To date, government programs with
regard to the environment and the workplace have not had an important independent
impact on technological innovation in the
ammonia industry.

This finding was established on the basis of
a thorough legal and engineering analysis of
ammonia technology. We did find that some
regulations -- boiler safety regulations -- have
existed for a long time and that, generally,
boilers and tanks at ammonia production sites are
safely designed. However, we do not believe that
it is accurate to conclude that those regulations
caused manufacturers to design safe containers.
Rather, it appears that commercial necessity and

179

professional architect/engineer standards contributed to those safe designs. The development
of all four facts -- safe design, commercial
necessity, professional standards, and boiler
safety regulations -- occurred roughly simultaneously, and the genesis for each was partially
dependent on each of the others.

If new emission standards are set considerably
below present levels, existing technology may
prove incapable of meeting the regulations without
major increases in production costs. We have not
investigated the tradeoffs between the environment
on the one hand and on food availability and
prices on the other. If new technology is
developed, the impact on production cost will be
mitigated in the longer run. We were not able to
observe this type of response in the case of
ammonia; it may appear more clearly in other products.

(3) Ammonia technology has proved capable
of adjusting to changed circumstances since
the time that ammonia was first synthesized.
A number of different feedstocks and basic
chemical approaches can be utilized in its
production.

This finding should be kept in mind as policy
is considered regarding special treatment for the
industry with respect to natural gas allocations.
It is true that existing facilities are highly
dependent upon natural gas as a feedstock and, to
a somewhat lesser extent, as a fuel. However, for
several decades coal played both roles, and ammonia
production may benefit from the current reawakening
of interest in the use of this plentiful material.
Moreover, research is being undertaken on methods
of stimulating plant growth which do not require
the application of traditional fertilizers. To
the extent that the relative scarcity of natural
gas is reflected in its price and availability,
research which seeks alternatives to the use of
natural gas as an input and ammonia as a fertilizer
will be stimulated. By the same token, favored
treatment for ammonia producers may inhibit the
search for alternatives and maintain reliance on
the dwindling supplies of natural gas further
into the future.

(4)　This study has shown that, in one case
at least, process technology has responded
to factor price changes in a direct way.
If this phenomenon can be verified more
generally, policy makers may have greater
confidence in the ability of the economy
to respond to changes in raw material prices
and availability dynamically through technical change in addition to its ability to
respond by reallocating supplies of inputs
among existing products and processes.

Despite its place in the conventional wisdom
of political economy, ours is the first study to
our knowledge which has demonstrated that industrial process technology responds in a meaningful
way to relative factor price changes at the process level of aggregation.　Further work is
desirable to test the general truth of this finding
and to see what factors influence the vigor with
which technologies respond.　If factor price
changes are the driving forces, what are the
resistances which establish the observed rates of
change in concert with the forces?　Ultimately,
one should like to be able to predict the response
of the "technology matrix" in an economy-wide
input/output model to price changes and other
factors.　That day lies considerably in the future.

Appendix: Economic and Technical Data on Ammonia Processes

A.1 INTRODUCTION

This appendix presents four sets of data: factor prices, output prices, process data on factor inputs, and individual plant information on processes and capacity. It includes evaluation of the quality of the raw data and discussion of the methods used to structure and analyze the final data series.

A.2 FACTOR PRICE DATA

The inputs of major interest to analysis of ammonia production include coal, fuel oil, natural gas, electricity, labor, catalysts and chemicals, and investment. In addition, costs of various modes of product transportation are important.

In this section we present price series for coal, natural gas, and electricity. Prices paid for catalysts and chemicals, as well as investment data, are presented in Tables A.4 to A.10. Unit labor cost data are available in standard sources and are not presented here; man-hours per unit of output appear in Section A.4. Transportation costs are discussed briefly in Section A.3.

Coal Prices

Coal price data are abundant. Three major government offices report mutually consistent current and historical price series. These figures are derived initially from detailed annual reports furnished by producers to the Department of the Interior's Bureau of Mines. Some small producers do not report directly to the Bureau of Mines.

Information from these producers is obtained
through state mine departments.

The Bureau of Mines tabulates the responses
from producers and state mine departments. Each
year the Bureau publishes its Minerals Yearbook
which includes a chapter on "Bituminous Coal and
Lignite." The many statistics used to compile the
yearbook appear to be the basis for the price
series reported by other agencies. The Department
of Commerce has a coal price series in its Survey
of Current Business. The Department of Labor
issues coal prices within its Wholesale Prices and
Wholesale Price Indexes. The prices listed in each
of the three publications are identical. The
Bureau of Mines reports prices on a state-by-state
basis as well as U.S. averages. A National Bureau
of Economic Research release, Report of the
Committee on Prices in the Bituminous Coal Industry
(1938), details how the Bureau of Mines determines
coal prices.

The Interstate Commerce Commission has
developed a schedule of the average cost of transporting coal. The brief title of the series is
"railroad freight charges per short ton." It
represents the average revenue received by Class I
railroads per net ton of bituminous coal as
reported to the Interstate Commerce Commission.
These figures, when added to the F.O.B. mine prices,
yield an excellent representation of delivered coal
costs.

Historical prices are available through past
yearbooks or in various historical surveys of
economic statistics. Figure A.1 is drawn from
data found in the U.S. Department of Commerce,
Bureau of the Census', Bicentennial Edition of
Historical Statistics of the United States, Colonial
Times to 1970. The "F.O.B. Mine" prices come from
series number 96 in the volume. "Delivered Price"
is F.O.B. price plus series number 99, the Interstate Commerce Commission's tally of railroad
freight charges.

Table A.1 is reproduced from the Federal
Energy Administration's pamphlet, Monthly Energy
Review. It shows prices that steam generating
electric power plants paid for coal delivered to
them in various parts of the country. These
figures indicate the movement of coal prices in
recent years.

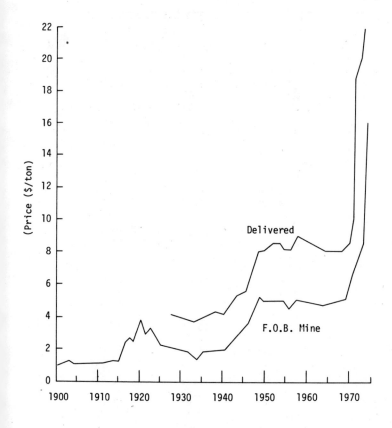

FIGURE A.1. Bituminous Coal Prices

Sources: "F.O.B. Mine" price - series number 96.
          "Delivered" price - series number 96 plus number 99.
          Both found in Historical Statistics of the United
          States, Colonial Times to 1970, U.S. Dept. of
          Commerce, Bureau of the Census.
          Post 1970 data is from Minerals Yearbook.

TABLE A.1
U.S. Average Delivered Prices of Coal at Utilities ($/ton)

		Contract	Spot
1973	January	8.09	9.91
	February	8.31	10.01
	March	8.42	10.07
	April	8.43	10.44
	May	8.51	10.24
	June	8.62	10.43
	July	8.44	10.40
	August	8.45	10.44
	September	8.71	10.67
	October	8.86	11.24
	November	9.13	12.05
	December	9.19	13.34
1974	January	9.83	17.02
	February	10.40	20.57
	March	10.63	22.54
	April	11.28	23.70
	May	11.80	24.21
	June	11.87	25.84
	July	12.05	27.99
	August	12.50	28.87
	September	12.89	30.64
	October	13.30	30.67
	November	14.16	31.95
	December	14.20	31.05
1975	January	14.57	28.12
	February	15.71	25.93
	March	15.68	25.02
	April	15.88	24.52
	May	16.45	23.78
	June	16.40	23.36
	July	16.06	22.35
	August	16.65	22.39
	September	16.76	22.46
	October	16.72	22.52
	November	16.79	22.50
	December	16.90	22.40

Source: Monthly Energy Review, Federal Energy Administration
(May 1976) page 70.

Natural Gas Prices

The major sources for historical natural gas
price data are the Federal Power Commission (F.P.C.),
the Bureau of Mines, and the gas industry itself.
The F.P.C. requires all firms which engage in
interstate sale of gas to report on their operations
and collects data on local sales from states and
individual firms. The commission has the most
extensive collection of statistics pertaining to
natural gas. Information about wellhead prices,
pipeline sales, and industrial prices is available
in various F.P.C. publications. Maps of the interstate pipeline system are also issued by the
agency.

The Bureau of Mines includes a chapter on
natural gas in its annual edition of Minerals
Yearbook. Production, consumption, pipeline transportation, and price statistics are reported. The
bureau receives information annually from oil and
gas producers, natural gas processing plants, pipeline companies, and gas utility companies. The
"Wellhead Price" series plotted in our Figure A.2
comes from successive issues of the Minerals
Yearbook.

The gas industry collects its own statistics.
The American Gas Association (A.G.A.) compiles
data provided by individual gas companies. The
A.G.A. receives annually the "Uniform Statistical
Report" from each company. This detailed ques-
tionnaire is also used by many companies to report
their operations to the financial community.
A.G.A. statistics are based on the Uniform Statistical Reports as well as other sources of information. The A.G.A. publishes Gas Facts each year.
This annual release, plus a summary document
published in 1956, Historical Statistics of the
Gas Industry, are the sources for "Industrial
Price (2)" and "Industrial Price (3)" also shown
in Figure A.2.

The first step in an attempt to chart the
price of natural gas is to obtain wellhead values.
Wellhead value is an accessible statistic which
serves as a base point from which other prices can
be calculated. Changes in all other prices
(including the prices paid by ammonia producers)
are paralleled by wellhead values.

The industrial price graphs (1), (2), and (3)
depict the prices faced by ammonia producers.
"Industrial Price (1)" is the industrial price

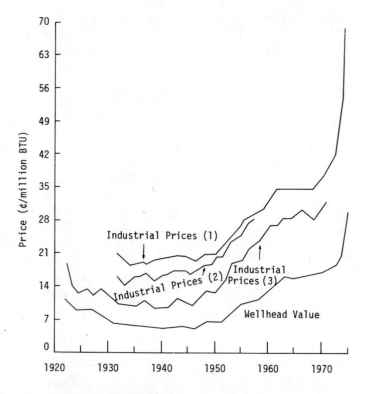

Price (¢/million BTU)

Industrial Prices (1)

Industrial Prices (2)

Industrial Prices (3)

Wellhead Value

FIGURE A.2. Natural Gas Prices

Sources: Wellhead Value - Historical Statistics of the United
  States, Department of Commerce, 1975; Series M 148
      Industrial Prices - (1) Industrial Average Prices, Ibid.,
                      Series S 17.
                  (2) Industrial Prices paid to Utilities.
  Total Gas Sales (including manufactured and mixed gas as
  well as natural gas) (1932-1954) Historical Statistics of
  the Gas Industry, American Gas Association, Tables 87 and
  107. (1955-1974) Gas Facts, Amer. Gas Assoc., 1974.
                  (3) Industrial Prices paid to Utilities.
  Natural Gas Only, (1932-1954) Historical Statistics of the
  Gas Industry, Tables 88 and 108; (1955-1964) Gas Facts,
  1965, Tables 84 and 104.

188

listed in Historical Statistics of the United
States. For this series the Department of Commerce
retrieved statistics from the annual editions of
the Minerals Yearbook. Industrial Prices (2) and
(3) are derived from statistics from the gas industry. Series (2) is the quotient of total gas
revenues divided by total gas sales. This yields
the average price in cents per million BTU.
"Industrial Price (3)" is the result of natural gas
revenues divided by natural gas sales. The
difference between the two series is that (2)
(total gas) includes manufactured and mixed gas as
well as natural gas. We notice that after about
1958 the two price series tend to merge. This is
due, presumably, to the decreased importance of
manufactured gas.

For our purposes, it is not enough to have a
series of average U.S. industrial prices for
natural gas since locational differences in prices
and availability play an important role in ammonia
plant location and process selection. It is
possible to construct a time series of industrial
prices for any particular state or region. The
F.P.C. lists revenues and sales differentiated by
class of use (e.g., industrial) and by state or
region. Thus, just as we computed series (3)
above, we can calculate an industrial price
series for any area. Typical values are shown in
Table A.2.

It must be noted that the natural gas industry
is regulated by federal and state authorities.
Natural gas prices therefore reflect not only the
normal market forces of supply and demand, but also
regulatory efforts. The federal regulation of
interstate natural gas sales was begun by the
F.P.C. in 1938 with the passage of the Natural Gas
Act. Regulation of wellhead prices of natural gas
destined for interstate sale was underway by 1954
when the Supreme Court upheld the F.P.C.'s
authority to regulate wellhead prices in Phillips
Petroleum Co. vs. Wisconsin, 347 U.S. 622 (1954).
During the sixties, individual rate decisions for
each producer were abandoned in favor of area-wide
maximum rates for each major producing location.

The effect of federal regulation on prices is
readily visible. (Note the appearance in Figure
A.2 of the virtual price freeze of 1960-68.)
Government policy also affected the availability of
natural gas in various parts of the country, but
availability is less visible than price in economic
statistics. Industrial natural gas users who buy

TABLE A.2
Prices for Industrial Use of Natural Gas by Region ($/million BTU)

	1950	1955	1960	1965	1970	1973	1974
New England	1.14	1.10	.97	.90	.95	1.25	1.66
Middle Atlantic	.46	.52	.59	.59	.63	.81	1.06
East North Central	.27	.34	.41	.41	.44	.57	.74
West North Central	.17	.24	.29	.30	.33	.44	.56
South Atlantic	.27	.33	.40	.41	.43	.58	.76
East South Central	.15	.20	.29	.28	.30	.44	.57
West South Central*	.08	.12	.16	.18	.21	.27	.45
Mountain	.14	.20	.24	.26	.29	.38	.52
Pacific	.19	.26	.32	.32	.34	.47	.67
Louisiana	.07	.10	.14	.16	.18	.26	.38
Texas	.08	.12	.15	.18	.21	.26	.47
United States	.21	.27	.33	.35	.38	.50	.68

Source: American Gas Association, Gas Facts, 1974.

This Table was constructed by converting the average gas prices paid in each region
(Table 94, Gas Facts) to industrial prices. The conversion factor used was the
ratio: National Average Industrial Price for each year.
         National Average Price

The statistics for the conversion factor came from Table 92, Gas Facts.

*West South Central includes Louisiana and Texas.

directly from interstate pipelines are exempt from
F.P.C. price regulation, but the regulatory
influence over the whole spectrum of natural gas
prices, as well as upon the availability of gas,
is important to our analysis of factor prices.

Apparently, new ammonia plants could secure
natural gas from interstate pipeline companies
until 1970 at least. Soon after that many pipeline companies were operating on a curtailment
basis, accepting no new customers at any price.
In order to evaluate the current natural gas price
situation we present Table A.3 reproduced from
the May 1976 issue of Monthly Energy Review, published by the Federal Energy Administration. The
series shows national average prices paid by
industrial users to pipeline companies. These
prices are not a full indication of the cost of
gas in times of shortage. Current spot intrastate
new gas sales at prices of almost two dollars per
thousand cubic feet attest to the opportunity cost
of today's natural gas.*

Electricity Prices

The price of electrical energy is reported by
many government agencies. The two major ones are
the U.S. Bureau of Labor Statistics and the F.P.C.
The Bureau of Labor Statistics issues electricity
prices monthly both in its Wholesale Prices and
Price Indexes and its Retail Prices and Indexes of
Fuels and Electricity. The F.P.C. publishes a
wealth of data, including information on each
individual power company. Each year the F.P.C.
prints a new edition of Typical Electric Bills
19XX; Residential, Commercial, and Industrial.
Within this release the utility rates in each city
of over 50,000 people are listed.

Ammonia producers are massive users of electric power. A typical 400-tpd reciprocating plant
used 7,500,000 KWH/month.** A modern 1000-tpd
centrifugal plant uses 600,000 KWH/month. Unfortunately the sources above do not report prices of
utility usage on such a large scale. The largest

---

*On July 27, 1976, the Federal Power Commission set
a wellhead price of $1.42 per thousand cubic feet
for interstate natural gas.

**625 KWH/ton x 400 tons/day x 30 days/month =
7,500,000 KWH/month.

TABLE A.3
Natural Gas Prices Reported by Major Interstate Pipeline
Companies for Direct Sales to Industrial Users
(¢/thousand cubic feet)

Date	Price
1974 January	48.1
February	49.8
March	50.8
April	49.3
May	49.3
June	50.8
July	52.5
August	55.2
September	54.7
October	56.3
November	58.7
December	60.3
1975 January	67.6
February	70.1
March	70.4
April	71.1
May	71.1
June	72.2
July	73.9
August	73.4
September	72.8
October	77.2
November	77.8
December	81.1
1976 January	87.5

Source: Monthly Energy Review (May 1976).

192

user classification in Typical Electric Bills is
only 200,000 KWH/month. Unit charges are reduced
as total usage increases; consequently the prices
for power which ammonia producers face are probably lower than the lowest price reported in
Typical Electric Bills. However, as can be seen
in Figure A.3, which shows prices over the years
for 30,000-, 60,000-, and 200,000-KWH/month usage,
the prices for each amount of service stay right
in step. Smaller users always pay a differential,
but over the years the differential has remained
rather constant, with prices fluctuating in
parallel. Thus we present the largest user series
(200,000 KWH/month) in Typical Electric Bills as
a first approximation to prices paid by ammonia
producers.

There are several problems with this series
for our purposes. First, as mentioned, the series
overstates the actual prices paid by producers.
Further, in many chemical industries, producers
generate on site, rather than purchase a portion
of their electrical needs. Industry sources
suggest that the extent of this activity in ammonia
facilities is small.

Price of Capital Variable

There are two aspects of the cost of capital.
First, there is the initial outlay, q, that must
be paid for each unit of capital. In this study,
q is set equal to the Chemical Engineering new
plant construction index. Since a capital good
delivers services for more than one year, the firm
apportions the total cost of the capital over its
lifetime. Each year the capital good depreciates
at a rate δ. We assume that depreciation is equal
to 5% of initial investment and therefore the
decline in value of a unit of capital good is δ
times q for any year.

However, there is also an opportunity cost for
the funds invested in the first year. If the investment is financed by internal funds this cost
is equal to the potential interest return forgone
due to the investment. If there is external
financing then this opportunity cost is the actual
interest payment made on the loan. In either case
this opportunity cost per unit of capital will be
equal to an interest rate, r, times q. We represent r by bond yields reported in Standard &
Poor's Trade and Securities Statistics, Security
Price Index Record, 1976 Edition.

193

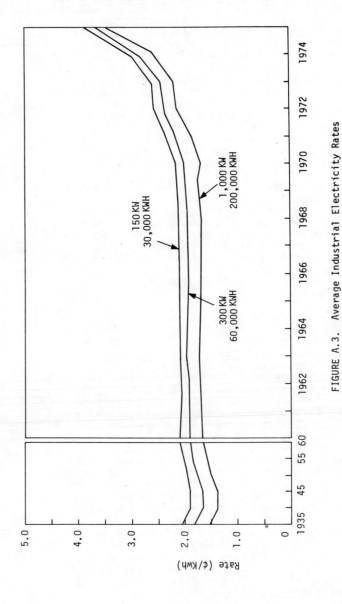

FIGURE A.3. Average Industrial Electricity Rates

Source: Typical Electric Bills 1975, Federal Power Commission, Wash., D.C.

194

Thus our price of capital variable is given by the equation:

$$P_k = (r + \delta) \; q \text{ where } \delta = .05.$$

A.3 AMMONIA PRICES AND TRANSPORTATION COSTS

Ammonia prices have been quoted weekly in the Oil, Paint, and Drug Reporter since at least 1923.* Over the years the exact specification of the commodity has varied slightly. At some points in time shipping charges have been included in the quoted price. The basic specification has always been: "Anhydrous Ammonia, Fertilizer Grade." Fertilizer Grade is distinguished from Refrigeration Grade, the latter being purer and a little more costly.

The Department of Labor reports on Anhydrous Ammonia prices in its publication, Wholesale Prices and Price Indexes. The publication is issued monthly with a year-end summary supplement. Until 1975 the figures always included shipping costs to points east of the Rockies except the East Coast; that is, shipping costs to points east of the Rockies not including the New England states, New York, Pennsylvania, Delaware, Maryland or New Jersey.

In April of 1957, the U.S.D.A., Agricultural Research Service published Statistical Bulletin #191, Statistics on Fertilizers and Liming Materials in the United States. Authors Mehring, Adams and Jacob computed "largely from published quotations in the Oil, Paint and Drug Reporter" annual ammonia prices from 1923 to 1952. The Oil Paint and Drug Reporter issued in 1971 a statistical compilation, Chemical Pricing Patterns, which summarized annual data from 1952 to 1970. From 1952 to 1968 these figures include "freight equalized" shipping costs from the nearest ammonia plant no matter which actually ships to him. Figure A.4 is drawn from the USDA bulletin figures, plus those from Chemical Price Patterns and figures computed directly from the Chemical Marketing Reporter.

---

*In 1972 the newspaper changed its name to the Chemical Marketing Reporter.

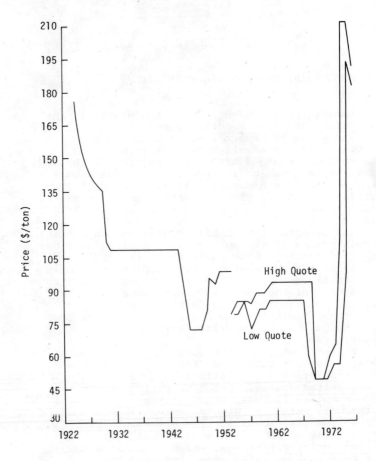

FIGURE A.4. Anhydrous Ammonia Prices

Sources: Statistics on Fertilizers and Liming Materials in the
United States, USDA Agricultural Bulletin #191; Chemical Pricing
Patterns; and Chemical Marketing Reporter.

Notes: 1952-1968, Freight Equalized East of the Rockies, except
East Coast.
Not freight equalized since 1969.
November 1973, Cost of Living Council removes price
controls.

Domestic final product ammonia is transported by four methods: pipeline, barge, railroad tank car, and tank trucks. Any one delivery may consist of several of these four modes of transportation as well as a storage stop at the plant, along the inland waterway, or at the pipeline terminal. Two large midwestern pipelines for ammonia were completed in 1962 and 1971. [1]

An idea of the comparative costs of the various modes of ammonia transportation can be gleaned from trade journal articles. Typical costs are as shown:

Mode	Freight Costs (¢/ton-mile)	
	1971	1968
Rail		
Single small car (11,000 gallons)	2.5	2.5-6.0
Single jumbo car (33,500 gallons)	1.2	2.0-3.0
Unit train	0.85	--
Truck	5.8	6.0
Pipeline	.73	--
Barge	.45	.54

A fifth mode of transportation, the ocean-going ammonia cargo ship, is employed to bring final product to the East Coast from plants located in Trinidad.

A.4 PROCESS DATA - FACTOR INPUTS
     AND INVESTMENT

Chapter 6 contains a discussion of the strengths and weaknesses of the engineering process coefficient data. From the data in the technical literature we have developed a set of time series of process coefficients which appear in Table 6.10. We have attempted to make data points comparable to each other by maintaining consistent interpretations of the variables. For instance, we have tried to maintain the concept of "battery limits" plant (which does not include storage facilities or administrative buildings) when estimating plant construction investment. Another important rule corrects all data points which include purchased steam inputs to an equivalent situation in which steam is generated on site. This requires that we

197

increase the natural gas fuel consumption by an
amount equal to that necessary to produce the
steam. (The small additional investment cost is
ignored.) We have therefore changed each data
point in which purchased steam is indicated to one
in which steam is omitted and 940,000 BTU of
natural gas per 1000 pounds of required steam is
added. The time series of process coefficients
attempts to present the factor input coefficients
associated with the most efficient plant size and
design available at each time period. The series,
with one exception, carries a coefficient through
time until a new observation appears in the
literature.

The one exception is that, for the period
1955-1964, we were not able to find articles giving
sufficient current information. We know that plant
efficiency was increasing (i.e., that factor input
requirements were falling), and we know that this
increased efficiency was achieved in large part
through increases in reforming pressures. [25]
The engineering literature reveals that the maximum
reforming pressures were 200 psi in the 1950s,
250 psi in 1960 and 1961, 300 psi in 1962 and 1963
and 400 psi in 1964. We used the 1954 data point
(45.2 million BTU gas, 676 KWH electricity, 1.47
man-hours labor, and capital intensity of 0.277)
and adjusted these coefficients downward according
to the efficiency gained by the increased
reformer pressures found in 1960-1964 plants. The
downward adjustment was based on Table 6.2 of
Chapter 6, which shows the theoretical reduction
in input requirements due to increasing reformer
pressure.

Thus, for instance, the derived data point for
1960 was determined as follows:

Gas Input = 22.3 (1954 feed requirement not affected by
increased pressures)

plus

22.9 (1954 fuel requirement)

x 125/128.2 (ratio of decreased fuel consumption
due to an increase in reformer pressure
from 200 to 250 psi)

= 44.6 million BTU

Electricity Input = 676 (1954 data point)

$$\qquad x\ 662/680 \text{ (ratio of decreased electricity consumption due to an increase in reformer pressure from 200 to 250 derived from Helscher's Thesis, Table 3.2, using straight line interpolation.)}$$

= 658 KWH

Labor Input = 1.47 (1954 data point) x $(400/200)^{.6}/400$

= 1.11 man hours

The $(400/200)^{.6}$ factor adjusts from a 200-tpd plant (1954) to a 400-tpd plant, assuming the 0.6 rule applies.

Investment = 6.66 (1957 \$ figure for 1954 plant)

$\qquad$ x $(400/200)^{.6}$ (adjusts for 400-tpd plant)

$\qquad$ x 5.2/5.4 (ratio of decreased investment cost due to an increase in reformer pressure from 200 to 250 psi, derived from Helscher's thesis, Table 3.2, using straight line interpolation)

= 9.72 million 1957 dollars.

Capital Intensity = $9.72/400^{.6}$ = .27

Tables A.4 - A.10 present much of the process information gleaned from the technical trade journals. In the tables, numbers in brackets refer to footnoted references which appear in the reference section.

TABLE A.4
Ammonia Production Inputs: Coke-Water Gas Process

Feedstock		Coke		
Capacity (tpd)		170	150	140
Date		1954	1949	1952
Reference		[2]	[3]	[4]
Inputs				
Nature	Units			
	(per ton of ammonia)			
VARIABLE				
Coke (Feedstock)	ton	1.27	1.4	1.4
Power	KWH	1,280		1,140
Coal (Fuel)	ton	--	1.25	--
Catalysts and Chemicals	dollar	--	.40	--
Labor	man-hour	4.36	2.5	--
Steam	1,000 pounds	12.4	--	10.84
FIXED				
Total investment	million 1957 dollars	--	12.7	--

200

TABLE A.5
Ammonia Production Inputs: Water-Gas Oxidation*

Feedstock		Coke
Capacity (tpd)		300
Date		1917
Reference		[5]
Inputs		
Nature	Units	
VARIABLE	(per ton of ammonia)	
Coke (Feedstock and Fuel)	ton	2
Coal	ton	1
Lignite	ton	2
Labor	man-hour	66
MAINTENANCE	% of investment	6
FIXED		
Total investment	million 1914 dollars	30

*Essentially the same as the coke-water gas process.

TABLE A.6
Ammonia Production Inputs: Partial Oxidation Process*

Inputs (per ton of ammonia)	Units	NG 100 1955 [6]	NG 200 1955 [6]	FO 100 1955 [6]	FO 200 1955 [6]	FO 660 1974 [7]	FO 1,000 1974 [8]b	FO 1,100 1974 [9]	FO 1,100 1974 [7]	FO 660 1966 [10]	FO 660 1966 [10]	FO 660 1966 [10]	FO 660 1966 [10]	FO 660 1966 [10]	Coal 100 1955 [6]	Coal 200 1955 [6]	Coal 660 1974 [7]b	Coal 1,000 1974 [8]b	Coal 1,000 1974 [8]c	Coal 1,100 1974 [10]a
VARIABLE																				
Natural Gas (Feedstock)	million BTU	27	27																	
Natural Gas (Fuel)	million BTU	22	22																	
Total	million BTU	49	49																	
Fuel Oil (Feedstock)	barrel			4.27	4.27	3.32			3.32											
Fuel Oil (Fuel)	barrel			2.54	2.54	2.38			2.38											
Total	barrel			6.81	6.81	5.70	6.0	6.0	5.70	5.76	5.68	5.62	5.91	6.76						
Coal (Feedstock)	ton														1.28	1.28	1.15			1.15
Coal (Fuel)	ton														.56	.56	1.05			1.05
Total	ton														1.84	1.84	2.20			2.20
Power	KWH	108	108	120	130	45	110	30	45	1,048	1,010	987	746	365	125	125	136	206	206	136
Boiler feed water	gallon	1,070	1,070	1,360	1,340	360	5,000	320	360	527	552	576	527	527	1,500	1,500	454	6,000	6,000	454
Cooling water	gallon	4,500	4,500	5,000	5,000	80,000	6,000	3,070	80,000	70,846	65,562	63,641	70,846	82,614	5,000	5,000	70,000	3,000	4,000	70,000
Steam	1,000 pounds	—	—	—	—	—	—	—	—	.75	.75	.75	.75	.75	—	—	.36	10	10	.36
Catalysts and chemicals	dollar	1.24	1.24	1.29	1.29	.36	.65	.30	.36	—	—	—	—	—	1.29	1.29	.72	.65	.80	.44
Labor	man-hour	1.8	.88	.88	.88	.37	.43	.12	.22	—	—	—	—	—	.88	.88		.70	.70	
MAINTENANCE	% investment/yr	4	4	4	4	2.5	—	—	2.5	—	—	—	—	—	4	4	2.5	—	—	2.5
FIXED																				
Battery Limits	million 1957 $					16.1			22.4								24.2			33.9
Storage	million 1957 $					1.2			1.2								1.2			1.2
Other	million 1957 $					5.5			7.6								8.2			10.6
Total	million 1957 $	4.5	7.7	4.1	8.1	22.7	35.3	36 to 42	31.2	9.4	9.9	10.4	9.4	9.4	4.8	8.3	33.6	36.4	39.4	45.7

*Dash indicates coefficient not specified.

a. Coal gasification process.
b. Texaco process.
c. Union Carbide process.

TABLE A.7
Ammonia Production Inputs: Steam Reforming, Reciprocating Compressors*

Feedstock: Natural Gas

Capacity		150	140	120	120	250	200	400	440	100	400	200	300	400	200	400
Date		1949	1952	1954	1954	1954	1955	1964	1964	1965	1965	1966	1966	1966	1968	1971
Reference		[3]	[4]	[11]	[11]	[2]	[12]	[13]	[14]	[15]	[16]	[17]	[17]	[17]	[18]	[19]

Inputs	Units															
Nature	(per ton of ammonia)															
VARIABLE																
Natural Gas (Feedstock)	million BTU	22		22.7	22.7	21.3	22.3		20.6		20.3					20.4
Natural Gas (Fuel)	million BTU	78		23.1	22.6	9.9	20.1		26.8		9.4					9.4
Total	million BTU	100	33.5	45.8	45.3	31.2	42.4	31.0	47.4a	30	29.8	30	30	30	29.5	29.8
Power	KWH	0	1,040	676	858	1,050	222e	700	0	600	624	625	625	625	600	624
Boiler feed water	gallon	--					930	300	7,000		553	550	550	550	554	553
Cooling water	gallon	--		79,000	77,000	176,500	21,000	63,000	--	--	40,680	50,000	50,000	50,000	40,700	1,476
Steam	1,000 pounds	.80	4.84	1.81	1.84	6.1	.92	1.10	--	1.00b	-.216				-.240	-.216
Catalysts and Chemicals	dollar						.89	.31	.6	2.4	1.70	.40	.40	.40	.57	.60
Labor	man-hour	.2		1.8	1.8	2.54			.6		.3	.60	.40	.26	.6	.24
MAINTENANCE	% of investment/yr	4	--	3	3	--	4	3½	4	5c	4	3	3	3	2.5	4
FIXED																
Battery Limits	million 1957 $							5.6	6.9	2.3					6.1	
Storage	million 1957 $							.9	6.8d	.6					--	
Other	million 1957 $							.6	1.0	.2					.4	
Total	million 1957 $	11.3	--	4.9	4.7	--	6.9	7.1	14.7	3.1	6.0	4.0	5.1	6.1		10.2

*Dash indicates coefficient not specified.

a. Includes 14.6 million BTU for electricity generation.
b. Includes water.
c. Includes taxes and insurance.
d. Includes utility generation.
e. Gas-engine-driven synthesis gas compressor; high pressure reforming.

203

TABLE A.8
Ammonia Production Inputs: Steam Reforming, Centrifugal Compressors*

Feedstock: Natural Gas

Inputs	Units	500	600	1,000	600	1,000	1,500	600	600	1,000	1,000	1,000	1,500	1,500
Capacity (tpd)		500	600	1,000	600	1,000	1,500	600	600	1,000	1,000	1,000	1,500	1,500
Date		1965	1965	1965	1966	1966	1966	1968	1968	1968	1968	1968	1968	1968
Reference		[15]	[16]	[15]	[17]	[17]	[17]	[18]	[20]	[18]	[21]	[20]	[20]	[18]
VARIABLE	(per ton of ammonia)													
Natural Gas (Feedstock)	million BTU		20.4						20.4			20.4	20.4	
Natural Gas (Fuel)	million BTU		14.0						13.8			13.8	13.8	
Total	million BTU	35.0	34.4	35.0	32.3	32	31.5	34	34.2	33.3	32	34.2	34.2	31.5
Power	KWH	50	4.3	50	30	30	30	25	4.6	21.6	24.4	4.6	4.6	20
Boiler feed water	gallon	--	494	--	550	550	550	494	494	494		494	494	494
Cooling water	gallon	--	67,200	--	50,000	50,000	50,000	67,200	67,200	67,200	3,400[b]	67,200	67,200	67,200
Steam	1,000 pounds		0					0	0	0	0	0	0	0
Catalysts and Chemicals	dollar	1.00	.58	1.00	.40	.40	.40	.73	.58	.73	.70	.58	.58	.73
Labor	man-hour	.69	.2	.42	.20	.12	.08	.2	.2	.12	.23	.12	.23	.08
MAINTENANCE	% of investment/yr	5[a]	4	5[a]	3	3	3	2.5	3	2.5	4	3	3	2.5
FIXED														
Battery Limits	million 1957 $	7.8	8.0	13.2				11.5		14.3	15.4	13.2	19.3	20.7
Storage	million 1957 $	1.9		3.3				--		--	--			--
Other	million 1957 $	1.5		2.5						2.3	3.3			2.1
Total	million 1957 $	11.2		19.0				12.5		15.7	--			22.8

*Dash indicates coefficient not specified.

a. Includes taxes and insurance.
b. Total water.

204

TABLE A.8 (continued)
Ammonia Production Inputs: Steam Reforming, Centrifugal Compressors*

Feedstock		Natural Gas												
Capacity		600	600	1,000	1,000	1,000	1,000	1,500	1,500	660	1,000	1,000	1,000	1,100
Date		1971	1971	1971	1971	1971	1971	1971	1971	1974	1974	1974	1974	1974
Reference		[19]	[19]	[19]	[19]	[19]	[22]	[19]	[19]	[7]	[9]	[9]	[23]	[7]
Inputs														
Nature	Units													
VARIABLE	(per ton of ammonia)													
Natural Gas (Feedstock)	million BTU	20.4	20.4	20.4	20.4	20.4	20.2	20.4	20.4	18.9	20.4	20.4		18.9
Natural Gas (Fuel)	million BTU	14.0	12.0	16.1	10.6	12.7	11.7	15.7	10.0	12.6	16.1	10.6		12.6
Total	million BTU	34.4	32.4	36.5	31.0	33.1	31.9	36.1	30.4	31.5	36.5	31.0	31.0	31.5
Power	KWH	15.6	15.6	15.3	15.3	15.3	15.3	15.2	15.2	30	15.5	15.5	20	30
Boiler feed water	gallon	528	552	527	560	527		528	562	545	560	560	4,000	545
Cooling water	gallon	2,448	1,872	2,966	1,858	2,448	3,000b	2,880	1,872	50,000	1,860	1,860	2,500	50,000
Steam	1,000 pounds	0	0	0	0									
Catalysts and Chemicals	dollar	.60	.60	.60	.60	.60	.80	.60	.60	.54	.70	.70	.90	.54
Labor	man-hour	.2	.2	.12	.12	.12	.10	.08	.08	.27	.10	.10	.33	.16
MAINTENANCE	% of investment/yr	3	3	3	3	3	2.5	3	3	2.5				2.5
FIXED														
Battery Limits	million 1957 $						13.6			12.1			25.8	17.0
Storage	million 1957 $									1.2				1.2
Other	million 1957 $						.5			4.3				5.7
Total	million 1957 $	11.8	12.0	14.7	15.4	15.2	14.1	20.2	21.2	17.6	28.5	30.3		23.9

*Dash indicates coefficient not specified.

a. Includes taxes and insurance.
b. Total water.

205

TABLE A.9
Ammonia Production Inputs: Steam Reforming of Naphtha*

Capacity (tpd)		440	660	1,000	1,000	1,100
Date		1964	1974	1974	1974	1974
Reference		[14]	[7]	[8]	[9]	[7]
Inputs						
Nature	Units					
VARIABLES	(per ton of ammonia)					
Naphtha (Feedstock)	ton	.530	.51	--	.554	.51
Naphtha (Fuel)	ton	0	.37	--	0	.37
Total	ton	.530	.88	.8	.554	.88
Natural Gas (Fuel)	million BTU				12.8	
Fuel Oil (Fuel)	ton	.587				
Power	KWH	0	45	22	21.6	45
Boiler feed water	gallon	690	454	4,000	670	454
Cooling	gallon	--	65,000	2,700	1,627	6,500
Steam	pound	--	--	--	--	--
Catalysts and Chemicals	dollar	1.10	.88	.92	.80	.72
Labor	man-hour	1.4	.72	.37	--	.18
MAINTENANCE						
FIXED						
Battery Limits	million 1957 $	8.0	13.9	28.6	--	19.4
Storage	million 1957 $	6.7a	1.2	--	--	1.2
Other	million 1957 $	.9	4.8	--	--	6.4
Total	million 1957 $	15.6	20.0	--	34.5	27.0

*Dash indicates coefficient not specified.

a. Includes utility generation.

206

TABLE A.10
Ammonia Production Inputs: Electrolysis

Feedstock		Water
Capacity (tpd)		10
Date		1925
Reference		[24]

Inputs		
Nature	Units	
VARIABLE	(per ton of ammonia)	
Power	KWH	12,000
Catalyst	dollar	--
Labor	man-hour	12
FIXED		
total investment	million 1925 $	0.7

A.5 PLANT DATA

We have compiled data on a plant-by-plant
basis for all ammonia plants which have ever been
in production in the U.S. through 1975. From
these data we have constructed a year by year
summary of ammonia capacity by process type, year
of opening and closing, ownership, and construction
contractor. The results are summarized in Figures
A.5 and A.6 (as well as Figure 6.2 of Section 6.2)
which show capacity by process. The data have been
put together from a number of sources. [26]

It should be noted that inaccuracies may
appear in the data. First, plant closings are
often not reported; hence, a plant may disappear
from view between two benchmark dates, but the
exact date of closing is unknown. Second, capacity
is in terms of rated capacity; actual output may
be quite different. Third, although our search
was intensive, some plants may have been missed.
Fourth, process data are sometimes inferred from
other information about the plant. However, we
believe that the inaccuracies are small relative to
the total capacity in any given year.

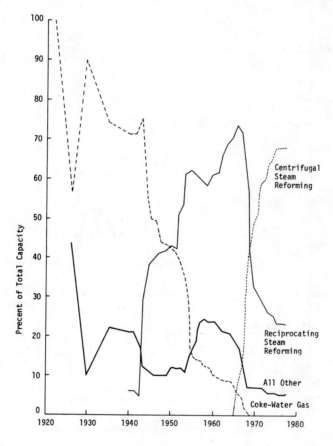

FIGURE A.5.   Distribution of Capacity of Major
Ammonia Processes

209

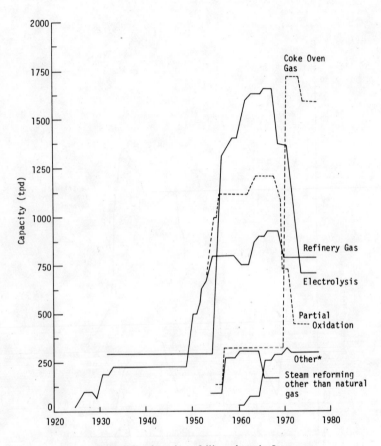

FIGURE A.6. Capacity of Minor Ammonia Processes

*Includes Refinery By-Product-Diffusion Separation, Waste Water Recovery, and Ethylene By-Product.

References

REFERENCES TO CHAPTER 1

1. "Facts and Figures for the Chemical
Industry," Chemical and Engineering News, 54:24
(June 7, 1976) pp. 33-48.
2. Haber, L.F., The Chemical Industry: 1900-
1930, Oxford University Press, London, 1971.
3. Slack, A.V. and G.R. James, Ammonia: Part I,
Marcel Dekker, Inc., New York, 1973.
4. Slack, A.V. and G.R. James, Ammonia: Part
II, Marcel Dekker, Inc., New York, 1974.
5. Sauchelli, V., Chemistry and Technology of
Fertilizers, Reinhold Publishing Corporation, New
York, 1960.
6. Sauchelli, V., Fertilizer Nitrogen, Reinhold
Publishing Corporation, New York, 1964.
7. Curtis, H.A., Fixed Nitrogen, Chemical
Catalog Company, New York, 1932.
8. Ernst, F.A., Fixation of Atmospheric
Nitrogen, Van Nostrand Co., New York, 1928.
9. Haynes, W., American Chemical Industry, six
volumes, Van Nostrand Co., New York, 1928.
10. Vancini, C.A., Synthesis of Ammonia, Mac-
Millan Press, Ltd., London, 1971.
11. Matasa, C. and E. Tonca, Basic Nitrogen
Compounds, Chemical Publishing Co., New York, 1973.
12. Development Planning and Research
Associates, Inc., "Economic Analysis of Effluent
Guidelines: Fertilizer Industry," NTIS, PB-241 315
(1974).
13. Haber, L. F., The Chemical Industry: 1900-
1930, Oxford University Press, London, 1971.
14. Ibid.
15. Martin, W.H., "Public Policy and Increased
Competition in the Synthetic Ammonia Industry,"
Quarterly Journal of Economics, Volume LXXIII, No. 3
(August 1959), pp. 373-392.

213

16. Markham, J.W., *The Fertilizer Industry: Study of an Imperfect Market*, Greenwood Press, New York, 1958 (reprinted 1969).

17. Annessen, R.J. and D. Gould, "Sour-Water Processing Turns Problem into Payout," *Chemical Engineering*, 78:7 (March 22, 1971) pp. 67-69.

REFERENCES TO CHAPTER 2

1. This point has been made by a number of economists, e.g., A.B. Atkinson and J.E. Stiglitz, "A New View of Technical Change," Economic Journal, Vol. LXXIX, No. 315 (September 1969), pp. 573-8; and P.M. Hohenberg, Chemicals in Western Europe: An Economic Study of Technical Change, Rand McNally and Company, Chicago, 1967.

2. N. Rosenberg, "Problems in the Economist's Conceptualization of Technological Innovation," History of Political Economy, Vol. 7, No. 4 (Winter 1975), pp. 456-81.

3. J. Schmookler, Invention and Economic Growth, Harvard University Press, Cambridge, 1966.

4. W.D. Nordhaus, "An Economic Theory of Technological Change," The American Economic Review, Vol. 59, No. 2 (May 1969), pp. 18-28.

5. For example, see H. Asher, Cost-Quantity Relationships in the Airframe Industry, RAND Corporation 12-291 (July 1, 1956).

6. K. Arrow, "The Economic Implications of Learning by Doing," Review of Economic Studies, Vol. XXIX (June 1962), pp. 155-73.

7. J. Schmookler, Invention and Economic Growth, Harvard University Press, Cambridge, 1966.

8. J.M. Utterback and W.J. Abernathy, "A Test of a Conceptual Model Linking Stages in Firms' Process and Product Innovation," BH5 74-23, Graduate School of Business Administration, Harvard University, Boston (November 1974).

9. C. Kennedy, "Induced Bias in Innovation and the Theory of Distribution," Economic Journal, Vol. 74 (September 1964), pp. 541-7.

10. M.J. Kamien and N.L. Schwartz, "Optimal Induced Technical Change," Econometrica, Vol. 36, No. 1 (January 1967), pp. 1-17.

11. R.R. Nelson and S.G. Winter, "Growth Theory from an Evolutionary Perspective: The Differential Productivity Puzzle," The American Economic Review, Vol. LXV, No. 2 (May 1975), pp. 338-44.

12. H.P. Binswanger, "A Microeconomic Approach to Induced Innovation," Economic Journal, Vol. 84, No. 336 (December 1974), pp. 940-58.

13. N. Rosenberg, "The Direction of Technological Change: Inducement Mechanisms and Focusing Devices," Economic Development and Cultural Change, Vol. 18 (October 1969), pp. 1-25.

14. C.T. Hill, et al., "A State of the Art Review of the Effects of Regulation on Technological Innovation in the Chemical and Allied Products Industries," Vol. II, Document PB-243 728/AS, NTIS, Springfield, Virginia, 1975.

15. M.I. Kamien and N.L. Schwartz, "Market Structure and Innovation: A Survey," Journal of Economic Literature, Vol. XIII, No. 1 (March 1975), pp. 1-37.

16. N. Rosenberg, "Science, Invention, and Economic Growth," Economic Journal, Vol. 84, No. 333 (March 1974), pp. 90-108.

17. F.T. Moore, "Economies of Scale: Some Statistical Evidence," Quarterly Journal of Economics, Vol. 73 (1959), pp. 232-45.

18. H.B. Chenery, "Engineering Production Functions," Quarterly Journal of Economics, Vol. 63 (1949), pp. 307-31.

19. S. Teitel, "Economies of Scale and Size of Plant: The Evidence and the Implications for the Developing Countries," Journal of Common Market Studies, Vol. XIII, Nos. 1 and 2 (1975), pp. 92-115.

20. J.G. Myers and L. Nakamura, "Energy and Pollution Effects on Productivity: Putty-Clay Approach," National Bureau of Economic Research, N.Y. (April 26, 1976). (Mimeo)

21. J.A. Finneran, N.J. Sweeney, and T.G. Hutchinson, "Startup Performance of Large Ammonia Plants," Chemical Engineering Progress, Vol. 64, No. 8 (August 1968), pp. 72-7.

22. A more complete discussion of the ideas raised in this paragraph may be found in the following: F.M. Scherer, et al., The Economies of Multi-Plant Operation, Harvard University Press, Cambridge, 1975; A.S. Manne (ed.), Investments for Capacity Expansion: Size, Location, and Time-Phasing, M.I.T. Press, Cambridge, 1967; and R.C. Levin, "Technical Change, Economies of Scale, and Market Structure," Yale University, New Haven, Ct. (1974),

Ph.D. dissertation. An early study of location and size of ammonia plants is by T. Vietorisz and A.S. Manne, "Chemical Processes, Plant Location and Economies of Scale," in Studies in Process Analysis: Economy-wide Production Capabilities, A.S. Manne and H.M. Markowitz (ed.), Cowles Foundation for Research in Economics Monograph No. 18, Wiley, New York (1963).

23. D.F. Rudd, "Modelling the Development of the Intermediate Chemicals Industry," in R.H. Day and T. Groves (ed.), Adaptive Economic Models, Academic Press, New York, 1975.

24. Anonymous, "Centrifugals Cut Ammonia Costs," Hydrocarbon Processing, Vol. 45, No. 5 (May 1966), pp. 179-80.

25. M.W. Kellog Co., "A Presentation of the High Capacity Single Train Ammonia Process," M.W. Kellog Co. (July 21, 1967). (Mimeo)

26. J.L. Marx, "Nitrogen Fixation: Prospects for Genetic Manipulation," Science, Vol. 196 (May 6, 1977), pp. 638-41.

27. T.P. Roth, "The Subjective Production Function: An Approach to its Determination," Engineering Economist, Vol. 17, No. 4 (Summer 1972), pp. 249-59.

28. N. Rosenberg, "Science, Invention and Economic Growth," Economic Journal, Vol. 84, No. 333 (March 1974), pp. 90-108.

REFERENCES TO CHAPTER 3

1. Occupational Health and Safety Act of 1970,
29 U.S.C. §651 et seq. (1970); Walsh-Healey Act, 41
U.S.C. §35 (1970).
2. Federal Water Pollution Control Act
Amendments of 1972, 33 U.S.C. §1251 et seq. (Supp.
III 1973); Clean Air Act Amendments of 1970, 42
U.S.C. §1857c (1970).
3. Conversation with Gary Meyers,
The Fertilizer Institute, in Washington, D.C.,
April 26, 1976.
4. Natural Gas Act §5, 15 U.S.C. §717d (1970):
"... the Commission shall determine the just and
reasonable rate ...." The Commission's jurisdiction
extends to "the transportation of natural gas for
resale for ultimate public consumption ... but shall
not apply ... to the local distribution of natural
gas." Natural Gas Act §1, 15 U.S.C. §717 (1970).
Thus, the FPC regulates the sale and transportation
of the gas from the wellhead to the distributor, but
not the actual use price. Until the energy crisis,
intrastate rates moved in a similar manner to interstate rates. Had they been significantly higher
users could move to other states.
5. Wall Street Journal, January 21, 1977, at
2, col. 3.
6. One of the worst disasters in the history
of American chemical manufacturing and handling
occurred in Texas City, Texas, on April 16-17, 1947.
576 persons died and property damage exceeded $67
million due to explosions and subsequent fires.
Chemical and Engineering News, April 4, 1976, at 201.
In the aftermath of that disaster, new restrictions
were placed on the handling and transportation of
ammonium nitrate by water. Interview with Harrie
Backes, Monsanto Corporation in St. Louis, Mo.,
May 12, 1976.

7. See, e.g., 46 C.F.R. §146.22-.30 (1976)
(Coast Guard, Department of Transportation (DOT),
Transportation and Storage of Explosives or Other
Dangerous Articles or Substances, and Combustible
Liquids on Board Vessels); 46 C.F.R. §146.23-100
(1976) (regulations pertaining to water transportation of nitric acid); 49 C.F.R. §170-89 (1976)
(regulations pertaining to motor and rail transportation of hazardous materials).
    8. See, e.g., Commoner, Improving Resources
Productivity: A Way to Supply the Growing World
Population, Ambio, Vol. 3, p. 137, 1974, for the
argument that American foreign aid should involve
shipping fertilizer, not food.
    9. Sherman Antitrust Act §§1-7, 15 U.S.C.
§§1-7 (1970); Federal Trade Commission Act §5, 15
U.S.C. §45 (1970); Clayton Anti-Trust Act §§3, 7,
15 U.S.C. §§14, 18 (1970); Robinson-Patman Anti-
discrimination Act §1, 15 U.S.C. §13 (1970).
    10. 16 U.S.C. §825(r) (1970).
    11. 15 U.S.C. §§79-79z(6) (1970).
    12. See, e.g., Securities Act of 1933, 15 U.S.C.
§§77a-77aa (1970); Securities Exchange Act of 1934,
15 U.S.C. §§78a-78jj (1970); Investment Company Act
of 1940, 15 U.S.C. §§80a-1 to 80a-52.
    13. 12 U.S.C. §§21 et seq (1970).
    14. See Federal Deposit Insurance Corporation
Act, 12 U.S.C. §§1811 et seq. (1970); Federal Home
Loan Bank Board Act (consolidated into the Housing
and Home Finance Agency by the Reorganization Plan
No. 3 of 1947, 61 Stat. 954, July 27, 1947);
Louisiana Banking Act, La. Rev. Stat. Ann. tit. 6,
§2 (West 1951).
    15. Land Grant Colleges Act, ch. 130, §§2 et
seq., 12 Stat. 503 (July 7, 1862), codified in 7
U.S.C. §§301 et seq (1970).
    16. 7 U.S.C. §§341 et seq. (1970); 7 U.S.C.
§361 (1970) (providing for agricultural experiment
stations).
    17. Agricultural Act of 1970, 7 U.S.C. §282a(d)
(1970).
    18. Agricultural Adjustment Act of 1938, 7
U.S.C. §§1281 et seq. (1970); Agricultural Act of
1949, 7 U.S.C. §§1421 et seq. (1970); see also
Federal Crop Insurance Act, 7 U.S.C. §§1501 et seq.
(1970); Agricultural Trade Development and
Assistance Act of 1954, 7 U.S.C. §§1691 et seq.
(1970).
    19. The movement that ultimately led to the
formation of the Tennessee Valley Authority (TVA)
originated in efforts to dispose of government

nitrate plants in Muscle Shoals, Alabama. After
several legislative attempts and Franklin D.
Roosevelt's inauguration, TVA was created to work
for flood control, soil conservation and other environmental goals. Indirectly, this left TVA with an
interest in fertilizer production. In addition, the
TVA rapidly undertook a program of fertilizer
experiments extending beyond the technical realm
into the commercial field. See 5 W. HAYNES, AMERICAN
CHEMICAL INDUSTRY 101-18 (1954). Announcement of the
coal gasification demonstration project was reported
in Chemical Engineering, June 6, 1977, p. 69.
          20. An important arm of TVA, for purposes of
this study, is the National Fertilizer Development
Center (NFDC), located in Muscle Shoals, Alabama.
Through research and development, the NFDC explores
new areas in fertilizer and fertilizer know-how
development. A recent entry to the technological
development field is the International Fertilizer
Development Center -- a private non-profit organization created to improve fertilizers and accompanying technology for developing countries. The
Center plans to cooperate with the NFDC in many
areas, saving time and money. INTERNATIONAL
FERTILIZER DEVELOPMENT CENTER, THE SHORT OUTLOOK
(1975).
          21. Int. Rev. Code of 1954, §38. For a brief
description of the working of the investment tax
credit, see Emory, The New Investment Credit, 3 Tax
Adv. 31 (1972); Woodward and Vincent, Investment
Influences of the Tax Credit Program, 18 Nat'l Tax
J. 272 (1965); Wilkson, The Investment Credit under
the Revenue Act of 1962, 42 Tex. L. Rev. 498 (1964).
          22. Under section 521 and 1381-88 of the
Internal Revenue Code of 1954, tax exemption is
extended to certain farmers' cooperatives, including
those organized "for the purpose of purchasing
supplies and equipment for the use of members ..."
Int. Rev. Code of 1954, §521. Amendments to the IRC,
enacted in 1962, impose tax treatment of either a
corporate tax on the cooperative or an individual
tax on its patrons. See Magnuson, All Farmers
Cooperatives and Other Corporations Operating on a
Cooperative Basis, 24 Coop. Acct. 52 (1971).
          23. Occupational Safety and Health Act of 1970
§5(a)(1), 29 U.S.C. §654(a)(1) (1970).
          24. 29 C.F.R. §1910.111 (1975), promulgated
39 Fed. Reg. 23502 (1974), as amended at 40 Fed.
Reg. 18426 (1975).
          25. 29 C.F.R. §1910.1000 (1975), promulgated
39 Fed. Reg. 23502 (1974).

26. 29 C.F.R. §1910.95 (occupational noise exposure) (1975), promulgated 39 Fed. Reg. 23502 (1974), as amended at 40 Fed. Reg. 23073 (1975).

27. 29 C.F.R. §§1910.21-.40 (1975), promulgated 39 Fed. Reg. 23502 (1974), as amended at 40 Fed. Reg. 18426 (1975).

28. Occupational Safety and Health Act of 1970 §6(a), 29 U.S.C. §655(a) (1970).

29. Walsh-Healey Act §1 et seq., 41 U.S.C. §§35-45 (1970).

30. 29 C.F.R. §1910.1000 (1975).

31. 40 Fed. Reg. 54684 (1975).

32. Id. at 54686.

33. Our conversations with industry experts supports this view, as does industry's formal response to the proposed OSHA standard. See Letter from Alied Chemical Company to Docket Officer, Docket No. H-053, OSHA, February 26, 1976 (cost of compliance would exceed $3 million); Letter from Smith Oil Co. to Docket Officer, Docket No. H-053, OSHA, June 3, 1976.

34. Letter from Allied Chemical Co. to Docket Officer, Docket No. H-053, OSHA, February 26, 1976.

35. Id.

36. 40 Fed. Reg. 54684-85 (1975); NIOSH, Criteria Document for Proposed Standard for Occupational Exposure to Ammonia (1975).

37. 41 Fed. Reg. 4943 (1976).

38. Letter from Grover Wrenn, Chief of Health Standards Development, OSHA, to David Newburger, June 24, 1976.

39. Interview with Grover Wrenn (see note 38 supra), in Washington, D.C., April 27, 1976; Interview with John Proctor, Senior Project Manager, Safety Standards Development, OSHA, in Washington, D.C., April 26, 1976.

40. American National Standards Institute K61.1-1966, Storage and Handling of Anhydrous Ammonia, codified at 29 C.F.R. §1910.111 (1975).

41. 41 Fed Reg. 17227 (1976).

42. Id. This is an example of the debate between performance and design characteristics.

43. Id.

44. Id. at 17253.

45. Id. at 17228.

46. 29 C.F.R. §§1910.35-.40, .94, .95, .105, .106, .109, .133, .134, .1000, .1002 (1975).

47. 29 C.F.R. §§1910.21 - .32, .101, .102, .104, .110, .166 - .171, .211 - .222, .251 - .254 (1975).

48. Occupational Safety and Health Act of 1970 §2(b), 29 U.S.C. §651(b) (1970).

49. Occupational Safety and Health Act of 1970 §4(a), 29 U.S.C. §653(a) (1970).

50. Wrenn and Proctor thought that there was a sufficient number of inspectors to inspect each employer covered under the Act once every 28 years. Note 39 supra.

51. We have suggested elsewhere that often a regulation is not what it appears to be, based on its written form. Hill, Greenberg, and Newburger, 2 A State of the Art Review of the Effects of Regulation on Technological Innovation in the Chemical and Allied Products Industries: Appendix A: The Legal Context of CAPI Regulation (1975). For instance, it is naive to think that a 55 mph speed limit means that everyone will drive at or less than 55 mph. The present speed limit appears to be causing most people to drive at or below 60 mph.

52. Occupational Safety and Health Act of 1970 §9(a), 29 U.S.C. §658(a) (1970).

53. Occupational Safety and Health Act of 1970 §10(a), 20 U.S.C. §659(a) (1970).

54. Occupational Safety and Health Act of 1970 §10(b), 29 U.S.C. §659(b) (1970).

55. Occupational Safety and Health Act of 1970 §17(b), (c), 29 U.S.C. §666(b), (c) (1970).

56. Occupational Safety and Health Act of 1970 §17(b), (c), (d), 29 U.S.C. §666(b), (c), (d) (1970).

57. Occupational Safety and Health Act of 1970 §17(a), 29 U.S.C. §666(a) (1970).

58. Occupational Safety and Health Act of 1970 §10(c), 29 U.S.C. §659(c) (1970).

59. Occupational Safety and Health Act of 1970 §11, 29 U.S.C. §660 (1970).

60. Interview with Proctor. Note 39 supra.

61. 29 U.S.C. §657(c) (1970).

62. 29 C.F.R. §1904 (1975).

63. Occupational Safety and Health Act of 1970 §8(f), 29 U.S.C. §657(f) (1970).

64. Occupational Safety and Health Act of 1970 §11(c), 29 U.S.C. §660(c) (1970).

65. For example, the fact that a citation has been issued, together with an explanation of the problem, is to be posted at or near the place where the violation occurred. Occupational Safety and Health Act of 1970 §9(b), 29 U.S.C. §658(b) (1970). Another example appears in the proposed ammonia exposure regulation which requires, among other things, training on the contents of the regulation. 29 C.F.R. §1910.1031 (j)(1)(ii)(F) (1975) [Proposed], 41 Fed. Reg. 54684, 54691 (1975).

66. S.3182, 94th Cong., 1st Sess. (1975).

67. Interview with John Proctor, note 39 supra.
68. 41 U.S.C. §§35-46, 45 Stat. 2036-39 (1936).
69. 41 U.S.C.§35 (1970).
70. 41 U.S.C. §35(e) (1970).
71. 41 U.S.C. §43 (1970).
72. Ruth Elkhorns Coals, Inc. v. Mitchell, 248
F. 2d 635 (D.C. Cir. 1957).
73. Id. at 638.
74. Id. at 639.
75. Interview with Robert Webber, Monsanto
Corp., St. Louis, MO., July 9, 1976.
76. See the reference to federal purchases of
superphosphate fertilizers in the early 1940s.
Mayo v. United States, 319 U.S. 441, 443 (1943).
77. Legislation: The Walsh-Healey Act: A
Congressional Gesture, 37 Colum. L. Rev. 102 (1937).
78. 41 U.S.C. §35(e) (1970).
79. 25 Fed Reg. 13809-25 (1960) (proposal for
41 C.F.R. §50-204 et seq).
80. Id. at 13809-10.
81. Id. at 13810.
82. Id. at 13823.
83. 28 Fed. Reg. 10524-40 (1963) (proposal for
41 C.F.R. §50-204).
84. 33 Fed. Reg. 14258-71 (1968) (proposal for
41 C.F.R. §§50-201, 50-204).
85. Id. at 14269.
86. Id.
87. 34 Fed. Reg. 788-96 (1969) (proposal for
41 C.F.R. §§50-201, 50-204.
88. Id. at 795.
89. Id. at 2207.
90. Id. at 7946-54.
91. Id. at 7953.
92. 41 C.F.R. §50-204.50 (1975).
93. 41 U.S.C. §36 (1970). A person who breaches
any of the provisions is blacklisted from entering
any new contracts with the federal government for
three years unless he receives special permission
from the Secretary of Labor for earlier contracts.
41 U.S.C. §37 (1970).
94. J. Page and M.W. O'Brien, Bitter Wages
(1973). Ray Davidson presents a similar story in
Perils on the Job, A Study of Hazards in the
Chemical Industry (1970). Davidson's book is an out-
growth of worker health and safety efforts by the
Oil, Chemical and Atomic Workers International
Union (AFL-CIO). See particularly Chapter 13 of the
latter work.
95. J. Page and M.W. O'Brien, supra note 94,
at 100.

96. See notes 53-57 supra and accompanying text.

97. See sources cited in note 94 supra.

98. Interview with Harrie Backes, Monsanto Corp., St. Louis, Mo., May 12, 1976; Interview with Grover Wrenn, supra note 39.

99. 470 Atlantic Avenue, Boston, Massachusetts 02210.

100. United Engineering Center, 345 East 47th Street, New York, New York 10017.

101. 29 C.F.R. §§1910.106, 1910.114 (1975).

102. NFPA No. 30-1969; 29 C.F.R. §1910.115 (1975).

103. See 29 C.F.R. §1910.106 (i)(2) (1975).

104. Cal. Stat. 1929, ch. 180, 181.

105. Cal. Labor Code §7621 (Deering 1971).

106. Id. §7622.

107. Id.

108. Id. §7680.

109. Id. §7681.

110. Id. §7682.

111. Id. §7683.

112. Id. §7650.

113. Id. §7686.

114. Id. §7687.

115. Id. §§7688, 7690.

116. Id. §§7691-92, 2250.

117. Cal. Stat. 1949, ch. 1530.

118. Cal. Stat. 1955, ch. 1784.

119. Cal. Stat. 1961, ch. 806.

120. Interview with Gerald Horn, Industrial Safety Engineer, California Industrial Safety Comm., July 20, 1976.

121. 49th Iowa Gen. Ass., ch. 97 (1971).

122. Id. §2(a).

123. Iowa Code Ann. §89.12 (1972), as amended, 58th Iowa Gen. Ass., ch. 105, §5 (1959).

124. Iowa Code Ann. §89.3 (1972). Other exemptions not relevant here.

125. Interview with J. Hanson, Boiler Supervisor, Iowa, July 20, 1976.

126. Iowa Code Ann. §89.2 (1972).

127. Id. §89.5.

128. Id. §§86.2-.6, 60th Iowa Gen Ass., ch. 93 (1963).

129. Id. §89.4 (1972), as amended, 49th Iowa Gen. Ass., ch. 97, §4 (1941); see also Op. Atty. Gen. (Iowa) Sept. 27, 1965: The Commissioner of Labor is not empowered to prescribe rules and regulations in respect to unfired pressure vessels

when such vessels do not contain water or steam, and
thus are not covered under Chapter 89.
130. Note 125 supra.
131. Iowa Code Ann. §89.9 (1972).
132. Iowa Code Ann. §89.10 (1972).
133. La. Rev. Stat. Ann. §23:540 (West 1964).
134. La. Rev. Stat. Ann. §23:536 (West 1964),
as amended, Acts 1966, No. 249, §1.
135. La. Rev. Stat. Ann. §23:537 (West 1964).
136. Id.
137. See La. Rev. Stat. Ann. §23:535 (West
1964).
138. La. Acts No. 264 §1 (1975).
139. La. Rev. Stat. Ann. §23:538 (West 1964).
140. Marcotte v. Ocean Acc. & Guaranty Corp.,
189 So. 2d 426 (La. App. 1966).
141. Olka. Laws 1919, ch. 146, §§1-7.
142. In 1968 the legislature established a
Bureau of Boiler Inspection, required a certificate
of operation, and exempted boilers inspected by
insurance companies from state inspection. Okla.
Laws 1968, ch. 234, §§1-3, codified at Okla. Stat.
tit. 40, §§148-50 (1975 Supp.).
143. Okla. Stat. tit. 40, §142 (1954).
144. Id. §145.
145. Id.
146. Okla. Stat. tit. 40, §112 (1954).
147. Okla. Stat. tit. 40, §143 (1954).
148. Okla. Stat. tit. 40, §147 (1954).
149. Acts 1937, 45th Leg. Sess., ch. 436 (Tex.).
150. See notes 133 and 143 supra.
151. Tex. Rev. Civ. Stat. art. 5221c, §3(1)
(1971).
152. Acts 1965, 59th Leg. Sess., ch. 440, §3a
(Tex.).
153. 1941 Op. Atty. Gen. No. 0-3459 (Tex.).
154. 1941 Op. Atty. Gen. No. 0-5139 (Tex.).
155. Tex. Rev. Civ. Stat. art. 5221c, §4 (1971).
156. Id.
157. Tex. Rev. Civ. Stat. art. 5221c, §6 (1971).
158. Id.
159. Tex. Rev. Civ. Stat. art. 5221c §2 (1971).
160. Tex. Rev. Civ. Stat. art. 5221c §4a (1971).
161. Tex. Rev. Civ. Stat. art. 5221c §5 (1971).
162. Tex. Rev. Civ. Stat. art. 5221c §2 (1971).
163. Tex. Rev. Civ. Stat. art. 5221c §4 (1971).
164. Tex. Rev. Civ. Stat. art. 5221c §13 (1971).
165. McFarland, Safety in Pressure Vessel
Design, 1970 Ammonia Plant Safety 473, 475 (1970).
166. New York Central R.R. v. White, 243 U.S.
188 (1917).

167. Cal. Stat., ch. 399, §796 (1911).
168. Iowa Laws, 35th Gen. Ass., ch. 147 (1913).
169. La. Acts No. 20 (1914).
170. Olka. Laws Art. 1, ch. 246 (1915).
171. Tex. Laws, ch. 113, §269 (1917).
172. The National Commission on State Workmen's Compensation Laws, The Report of the National Commission on State Workmen's Compensation Laws (1972) (hereinafter cited as Report); the National Commission on State Workmen's Compensation Laws, Compendium on Workmen's Compensation (1972); I, II, III Supplemental Studies for the National Commission on State Workmen's Compensation Laws (M. Berkowitz ed. 1973).
173. Iowa Laws, 52d Gen. Ass., ch. 71 (1947).
174. La. Acts No. 532 (1952).
175. Okla. Laws No. 427 (1953).
176. Tex. Laws ch. 133, §176 (1947).
177. Report at 18.
178. For cases which demonstrate that liberal interpretation, see Fidelity and Casualty Co. v. Industrial Acc. Comm., 177 Cal. 614, 171 P. 429 (1918); Collins v. Armour & Co., 11 So. 2d 621 (La. App. 1943); Bryant v. Beson, 53 Okla. 57, 4 P.2d 1061 (1931); Commercial Standard Ins. Co. v. Noack, 45 S.W. 2d 798 (Tex. Civ. App. 1931), rev'd on other grounds, 62 S.W.2d 72 (Tex. Civ. App. 1933); c.f. Black v. Creston Auto Co., 225 Iowa 671, 281 N.W. 189 (exposure to fumes unexpectedly at the workplace was an "accident," suggesting that fumes which are known to exist cannot cause an "accident.") Of course, a worker may wish to avoid the accident label in order to bring an action under ordinary tort law, in which case he may enhance his chances for a bigger award. See, e.g., Industrial Comm. v. Roth, 98 Ohio St. 34, 120 N.E. 172 (1918).
179. See Report at 18-20.
180. Report at 24.
181. Pub. L. No. 90-500, §2 (Oct. 18, 1972) (codified at 33 U.S.C. §§1251 et seq. (Supp. III 1973)).
182. Hooker Chem. & Plastics Corp. v. Train, 537 F. 2d 620, 623 (2d Cir. 1976).
183. Comment, The Federal Water Pollution Control Act Amendments of 1972, 14 B.C. Com. & Ind. L. Rev. 672, 687-90 (1973).
184. 33 U.S.C. §1251(a)(1) (Supp. III 1973).
185. 33 U.S.C. §1316(b)(1)(A) (Supp. III 1973).
186. Id.
187. The First Congress appropriated funds for the construction of a lighthouse at the entrance to

226

Chesapeake Bay. There were many subsequent measures
which dealt with specific pollution problems. See,
e.g., Rivers and Harbors Act of 1899, 33 U.S.C.
§407 (1970) (originally enacted as Act of Feb. 28,
1899, ch. 425, 13, 30 Stat. 1152); Act of Aug. 21,
1916, ch. 360, §3, 39 Stat. 518; Oil Pollution Act
of 1924, ch. 260 §3 Stat. 604 (repealed by Act of
April 3, 1970, Pub. L. No. 91-224, §108, 84 Stat.
91).
188. Act of June 30, 1948, ch. 758, 62 Stat.
1155 [hereinafter 1948 Act].
189. Berry, The Evolution of the Enforcement
Provisions of the Federal Water Pollution Control
Act: A Study of the Difficulty in Developing
Effective Legislation, 68 Mich. L. Rev. 1103, 1104
(1970).
190. 1948 Act §10(e).
191. Berry, supra note 189, at 1106-07; Davis
and Glasser, The Discharge Permit Program under the
Federal Water Pollution Control Act of 1972 --
Improvement of Water Quality through the Regulation
of Discharges from Industrial Facilities, 2 Ford.
Urb. L. J. 179, 227 (1973).
192. Act of July 9, 1956, ch. 518, 70 Stat.
498 [hereinafter 1956 Act].
193. 1956 Act §8(d), as amended, in 1966 (act
inserted built-in six month delay); 1956 Act §8(f)
(before abatement action could be filed, Secretary
of HEW must secure consent of either the state in
which pollution originated or the states in which
the health and welfare of persons were affected).
194. Federal Water Pollution Control
Amendments of 1961, Pub. L. No. 87-88, 75 Stat.
204 [hereinafter 1961 Act].
195. 1961 Act §§8a, 9e; see Berry, supra note
189, at 1113.
196. See S. Rep. No. 353, 87th Cong., 1st Sess.
4 (1961).
197. 1961 Act §8a.
198. Act of Oct. 2, 1965, Pub. L. No. 89-234,
79 Stat. 903 [hereinafter 1965 Act].
199. 1965 Act §5(a).
200. Id.
201. Davis and Glasser, supra note 191, at 192-93.
202. See, e.g., Hines, Controlling Industrial
Water Pollution: Color the Problem Green, 9 B.C.
Ind. & Comm. L. Rev. 553, 570 (1968).
203. 1965 Act §5(a).
204. Act of Nov. 3, 1966, Pub. L. No. 89-753,
80 Stat. 1246 [hereinafter 1966 Act].
205. 1966 Act §208(b).

206. See Berry, *supra* note 189, at 1119.
207. Act of April 3, 1970, Pub. L. No. 91-224,
84 Stat. 91 [hereinafter 1970 Act] .
208. 1970 Act §103.
209. Cf. Comment, *supra* note 183, at 678.
210. 33 U.S.C. §1316(b)(1)(B) (Supp. III 1973).
211. Id. §1311(b)(1)(A).
212. Id. §1311(b)(2)(A).
213. Id.
214. Id. §1311(a).
215. Id.; id. §1342(k); 39 Fed. Reg. 12835
(1974).
216. Id. §§1341(a)(1), 1342.
217. Comment, *supra* note 183, at 696.
218. E.I. DuPont de Nemours & Co. v. Train,
528 F. 2d 1136, 1139 (4th Cir. 1975), aff'd, 45
U.S.L.W. 4212 (Feb. 23, 1977).
219. Id.; CPC International, Inc. v. Train,
515 F.2d 1032 (8th Cir. 1975); American Iron &
Steel Inst. v. Environmental Protection Agency, 526
F. 2d 1027 (3d Cir. 1975); American Meat Inst. v.
Environmental Protection Agency, 526 F.2d 442 (7th
Cir. 1975); Hooker Chem. & Plastics Corp. v. Train,
537 F.2d 620 (2d Cir. 1976).
220. E.I. DuPont de Nemours & Co. v. Train 45
U.S.L.W. 4212, 4213 (Feb. 23, 1977).
221. Hooker Chem. & Plastics Corp. v. Train,
537 F.2d 620, 623 (2nd Cir. 1976).
222. 39 Fed. Reg. 12836 (1974).
223. See cases cited in notes 218 and 219
supra.
224. E.I. Dupont de Nemours & Co. v. Train,
45 U.S.L.W. 4212, 4213 (Feb. 23, 1977).
225. 39 Fed. Reg. 12832, 12836 (1974).
226. Id.
227. Id. at 33853
228. Id.
229. Id.
230. Hooker Chem. & Plastics Corp. v. Train,
537 F.2d 620 (2d Cir. 1976); American Iron & Steel
Inst. v. Environmental Protection Agency, 526 F.2d
1027 (3d Cir. 1975); American Meat Inst. v. Environmental Protection Agency, 526 F.2d 442 (7th Cir.
1975); E.I. DuPont de Nemours & Co. v. Train, 528
F.2d 1136 (4th Cir. 1975), aff'd on other grnds, 45
U.S.L.W. 4212 (Feb. 23, 1977).
231. See EPA Development Document for Effluent
Limitation Guidelines and New Source Performance
Standards for the Basic Fertilizer Chemicals (1974).
232. See note 195 supra.
233. Development Document, *supra* note 231, at 143.

228

234. Id.
235. Id.
236. 38 Fed. Reg. 33853 (1973).
237. Id.
238. Development Document, supra note 231, at
81.
239. Id. at 81-82
240. 39 Fed. Reg. 12838 (1974).
241. Development Document, supra note 231, at
149.
242. 39 Fed. Reg. 12838 (1974).
243. Id. at 12834-35.
244. 40 Fed. Reg. 26275 (1975).
245. Id.
246. 38 Fed. Reg. 33852, 33858 (1973).
247. 39 Fed. Reg. 12834 (1974) (Comment 13).
248. Id.
249. 38 Fed. Reg. 33854 (1973).
250. Id.
251. Id.
252. See note 247 supra .
253. 39 Fed. Reg. 12834 (1974).
254. Id. at 12833-34.
255. Development Document, supra note 231, at
148.
256. 39 Fed. Reg. 12833 (1974).
257. Interview with E.C. Martin, Standards
Development Section, Environmental Protection
Agency, in Washington, D.C. (Nov. 23, 1976).
258. Docket No. 74-1755 (6th Cir. 1976).
259. Interview with Karl Johnson, Director,
Member Services, the Fertilizer Institute, in
Washington, D.C. (April 26, 1976).
260. Id.
261. 41 Fed. Reg. 2387 (1976). Since the
settlement left further decisions to be made by EPA,
the matter continued on the Court of Appeals docket
until July 1978. The court then issued a two page
order remanding to EPA standard setting for nitric
acid -- the only remaining subject to be resolved.
The court observed that EPA had decided that setting
a nitric acid standard was a low priority for the
agency. Vistron Corp. v. Costle, Docket No. 74-1755
(remanded, July 11, 1978).
262. 40 C.F.R. §§418.50 - .56 (1975).
263. 41 Fed. Reg. 2387 (1976).
264. Id.
265. Id.
266. Development Document, supra note 231, at
80.
267. Id.; 38 Fed. Reg. 33852, 33853 (1973).

268. 41 Fed. Reg. 2387 (1976).

269. 39 Fed. Reg. 12833, 12834 (1974).

270. FMC Corp. v. Train, 539 F.2d 973, 981 (4th Cir. 1976).

271. Development Document, supra note 231, 10-31.

272. See, e.g., id. at 148.

273. See, e.g., Report from Farmland Industries, Inc., filed with Iowa Dept. of Environmental Quality, May 1, 1973.

274. Ch. 360, 69 Stat. 322 (1955), as amended, Act of June 8, 1960, Pub. L. No. 86-493, 44 Stat. 162, and Act of Oct. 9, 1962, Pub. L. No. 87-761, 76 Stat. 760.

275. Clean Air Act of 1963, Pub. L. No. 88-206, 77 Stat. 392, as amended , Air Quality Act of 1967, Pub. L. No. 90-148, 81 Stat. 485, and Clean Air Amendments of 1970, Pub. L. No. 91-604, 84 Stat. 1676.

276. See H.R. Rep. No. 1146, 91st Cong., 2d Sess. 1 (1970); S. Rep. No. 1196, 91st Cong., 2d Sess. 5 (1970); Note, Enforcement of the Clean Air Amendments of 1970, 10 Urb. L. Ann. 297, 298 (1975).

277. 42 U.S.C. §1857(c) - 6(a) (1970).

278. Id. §1857c - 7(b).

279. Id. §1857c - 8(a)(1).

280. Id. §1857c - 8(a)(2).

281. Id. §1857c - 4; see Note, supra note 276, at 298n.6.

282. 42 U.S.C. §1857c - 5(a)(1) (1970).

283. Id. §1857d - 1, as amended, (Supp. IV, 1974).

284. The Clean Air Amendments provided that the primary standard should be attained as "expeditiously as practical" but (subject to allowable variances) no later than three years from the date of the approval of the implementation plan. 42 U.S.C. §1857c - 5(a)(2)(A)(i) (1970). The secondary standard is to be met in a "reasonable time." Id. §1857c - 5(a)(2)(A)(ii).

285. Id. §1857c - 5(a)(2). (Approval is conditioned on compliance with the guidelines outlined in the statute. Id.)

286. Id. §1857c - 5, as amended, (Supp. IV, 1974).

287. 421 U.S. 60 (1975).

288. 44 U.S.L.W. 5060 (June 25, 1976).

289. See 40 C.F.R. §§50.6-.7 (1976) (primary and secondary ambient air quality standards for particulate matter). Id. §50.10 (primary and secondary ambient air quality standard for hydrocarbons). Id. §50.11 (primary and secondary

ambient air quality standard for nitrogen
dioxide).
     290. 42 U.S.C. §1857c - 2 (1970).
     291. Id. §1857c - 5(f).
     292. See Chapter 5 infra.
     293. Id. §1857c - 6. In addition to esta-
blishing federal performance standards for new
sources, the Amendments direct the administrator to
prescribe emission standards for any existing
source of any air pollutant which would be subject
to a performance standard if it were a new source.
Id.
     294. See 40 C.F.R. §§60.70-.74 (1976).
     295. See Chapter 5 infra.
     296. Minnesota Nitric Acid Plant Standards,
ch. 16: APC 16, reported in Environment Reporter:
State Air Laws, 416:0761 (1976); Texas Regulation
VII: Control of Air Pollution from Nitrogen Compounds, Rule 702, reported in Environment
Reporter: State Air Laws, 521:0681 (1976).
     297. See Environment Reporter, State Air
Laws 351:0509 (1976); see also Missouri Regulation
S-V, Restriction of Emission of Particulate Matter
from Industrial Processes, adopted Feb. 24, 1971,
amended, Jan. 18, 1972.
     298. Primary and secondary ambient air quality
standards were not promulgated until 1975. See 40
Fed. Reg. 7043 (1975). The Amendments direct the
administrator to revise the list of pollutants for
which air quality standards are appropriate. The
grounds for inclusion on the list are (1) the
pollutant adversely affects the public health or
welfare; (2) the presence of the pollutant in the
ambient air results from numerous or diverse mobile
or stationary sources; and (3) air quality criteria
for the pollutant had not been issued before
December 31, 1970. See 42 U.S.C. §1857c - 3 (1970).
     299. Public Law No. 95-95, 91 Stat. 685 (1977).
     300. For a general discussion of those
amendments, see Raffle, The New Clean Air Act --
Getting Clean and Staying Clean, Environment
Reporter Monograph No. 26, May 19, 1978.

REFERENCES TO CHAPTER 4

1. Slack, A.V. and G.R. James, Ammonia:
Part I, Marcel Dekker, Inc., New York, 1973, p. 37.
2. Ibid.
3. Haber, L.F., The Chemical Industry: 1900-
1930, Oxford University Press, London (1971).
4. Slack and James, Ammonia: Part I.
5. Sauchelli, V. (ed.), Fertilizer Nitrogen:
Its Chemistry and Technology, ACS Monograph Series
No. 161, Reinhold Publishing Corporation, New York,
1964.
6. Haynes, W., American Chemical Industry,
Volume II, Van Nostrand Co., New York, 1948, p. 87.
7. Sauchelli, Fertilizer Nitrogen.
8. Curtis, H.A., Fixed Nitrogen, Chemical
Catalog Company, New York, 1932.
9. Slack and James, Ammonia: Part I.
10. Sauchelli, Fertilizer Nitrogen.
11. Ibid.
12. Slack, A.V. and G.R. James, Ammonia: Part
II, Marcel Dekker, Inc., New York, 1974.
13. Sauchelli, Fertilizer Nitrogen.
14. Ernst, F.A., Fixation of Atmospheric
Nitrogen, Van Nostrand Co., New York, 1928.
15. Reynolds, P.W., "The Manufacture of
Ammonia," Fertilizer Society Proceedings, 89,
pp. 2-27 (November 25, 1965).
16. Ernst, F.A., "Ammonia Production and
Conversion Costs," Transactions of the American
Institute of Chemical Engineers, 17: Part 1,
pp. 1-26 (December 3, 1925).
17. Sauchelli, Fertilizer Nitrogen.
18. Anonymous, "Ammonia's New World: More
Plant, Less Crew," Business Week, pp. 134-139
(November 13, 1965).

19. M.W. Kellogg Co., "A Presentation of the High Capacity Single Train Ammonia Process," M.W. Kellogg Co. (July 21, 1967). (Mimeo)

20. Slack and James, Ammonia: Part II.

21. Piombino, A.J., "Ammonia - Where the Action Is," Chemical Week, 97:11, pp. 51-69 (September 11, 1965).

22. Anonymous, "Bigger Ammonia Plants Suggest Compressor Shift," Chemical Engineering, 70:17, pp. 88-90 (August 19, 1963).

23. Levin, R.C., "Technical Change, Economies of Scale, and Market Structure," Yale University, New Haven, Connecticut, 1974. (Unpublished Ph.D. Thesis).

24. Scherer, F.M., Industrial Market Structure and Economic Performance, Rand McNally College Publishing Co., Chicago, 1970.

25. Vietorisz, T. and A.S. Manne, "Chemical Processes, Plant Location and Economies of Scale," Studies in Process Analysis: Economy-wide Production Capabilities, edited by A. S. Manne and H.M. Markowitz, Cowles Foundation for Research Economics Monograph No. 18, Wiley, New York (1963).

26. Anonymous, "Kellogg's Single-Train Large Ammonia Plants Achieve Lowest Costs," Chemical Engineering, pp. 112-117, November 20, 1967.

27. M.W. Kellogg Co., "Single Train Ammonia Process."

28. McMichael, P., "Cheaper Ammonia with Off-Peak Electric Power," Chemical and Metallurgical Engineering, 37:8, pp. 484-487 (August, 1930).

29. Rosenstein, L., "Why Shell Built Its Ammonia Plant in California," Chemical and Metallurgical Engineering, 38:11, pp. 636-637 (November, 1931).

30. Annessen, R.J. and D. Gould, "Sour-Water Processing Turns Problem into Payout," Chemical Engineering, 78-7, pp. 67-69 (March 22, 1971).

REFERENCES TO CHAPTER 5

1. Finneran, J.A. and P.H. Whelchel, "Recovery and Reuse of Aqueous Effluent from a Modern Ammonia Plant," AIChE Workshop on Industrial Process Design for Water Pollution Control, San Francisco, CA., March 31-April 2, 1970, p. 109.
2. Ibid.
3. Ibid.
4. Romero, C.J. and F.H. Yocum, "Treatment of Ammonia Plant Process Condensate," Proceedings of the Fertilizer Institute Environmental Symposium, New Orleans, January 13-16, 1976, p. ·45.
5. Quartulli, O.J., "Review of Methods for Handling Ammonia Plant Process Condensate," AIChE Workshop on Industrial Process Design for Water Pollution Control, San Francisco, CA., March 31-April 2, 1970, p. 25.
6. Environmental Protection Agency, Inorganic Fertilizer and Phosphate Mining Industries - Water Pollution and Control, Report #12020 FPD 09/71, September 1971.
7. Finneran and Whelchel, op. cit.
8. Environmental Protection Agency, op. cit., p. 132.
9. Ibid., P. 141.
10. Pylant, H.S., "A Decade of Progress in Ammonia Technology," Oil and Gas Journal, 62:5, pp. 68-74 (February 3, 1964).
11. Environmental Protection Agency, op. cit., p. 93.
12. Ibid., p. 95.
13. Ibid.
14. Bingham, E.C., "Control of Air Pollution from Fertilizer Production," Chapter 26 in Industrial Air Pollution Control, K. Noll and J. Duncan, eds., Ann Arbor Science Publishers, 1973.

15. Finneran, J.H., M.W. Kellogg, Co., Houston, Tx. Personal communication to C.T. Hill, June 11, 1976.

16. Environmental Protection Agency, op. cit., p. 140.

17. Ibid., p. 149.

18. Development Planning and Research Associates, Inc., "Economic Analysis of Effluent Guidelines - Fertilizer Industry," Report to EPA. January 1974. Available from NTIS as PB-241-315.

19. Quartulli, op. cit.

20. Romero and Yocum, op. cit.

21. Calloway, J.A., A.K. Schwartz, Jr., and R.G. Thompson, "An Integrated Power Process Model of Water Use and Waste Water Treatment in Ammonia Production," Report to NSF, University of Houston, Houston, Texas, February 1974. Available from NTIS as PB-237-219.

22. Brill, W.J., "Biological Nitrogen Fixation," Scientific American, 236:3, p. 68, March, 1977.

23. Haynes, W., The American Chemical Industry, Van Nostrand, New York, 1948.

24. Commoner, B., The Closing Circle, Knopf, New York, 1971.

25. Richards, B., "Nitrogen Fertilizers Seen as Threat to Earth's Ozone Shield," The Washington Post, page A3, March 17, 1977.

REFERENCES TO CHAPTER 6

1. Anonymous, "Centrifugals Cut Ammonia Costs," Hydrocarbon Processing, 45:5, pp. 179-180 (May, 1966).

2. This section is largely based on T.M. Helscher, Process Innovation in the Manufacture of Ammonia: A Case Study of Steam-Reforming Natural Gas, master's thesis, Sever Institute of Technology, Washington University, St. Louis, 1977.

3. Sinfelt, John H., "Heterogeneous Catalysis: Some Recent Developments," Science, 195:4279, pp. 641-46 (February 18, 1977).

4. Mayland, B.O., E.A. Comley, and J.C. Reynolds, "Ammonia Synthesis Gas," Canadian Chemical Processing, pp. 22-29 (August, 1954).

5. Based on Quartulli, O.J., "Check List for High Pressure Reforming," Hydrocarbon Processing, 44:4, pp. 151-162 (April, 1965).

6. Based on Mayland, et al., op. cit.

7. See Quartulli, op. cit.

8. Pylant, H.S., "A Decade of Progress in Ammonia Technology," Oil and Gas Journal, 62:5, pp. 68-74 (February 3, 1964).

9. Based on Quartulli, op. cit.

10. Quartulli, O.J., and W. Turner, "Economic and Technological Factors in Giant Ammonia Plant Design," Nitrogen, No. 80, pp. 28-34 (1972).

11. Based on Quartulli, (1965), op. cit.

12. Slack, A.V. and G.R. James, Ammonia: Part II, Marcel Dekker, Inc., New York, 1974.

13. Based on Quartulli (1972), op. cit.

14. Axelrod, L., R.E. Daze, and H.P. Wickham, "The Large Plant Concept," Chemical Engineering Progress, 64:7, pp. 17-25 (July, 1968).

15. Levin, R.C., Technical Change, Economies of Scale, and Market Structure, Yale University, New Haven, Connecticut, 1974. (unpublished Ph.D. Thesis).

16. Finneran, J.A. and T.A. Czuppon, "Economics of Production and Consumption of Ammonia," M.W. Kellogg Company, Houston, Texas (November 23, 1973). (Mimeo)

17. See Pylant, op. cit., and Labrow, S., "Influence on Ammonia Synthesis of Computer Design," Chemical and Processing Engineering, 49:1, 55-57 (January, 1968). Manning, W.R.D. and S. Labrow, High Pressure Engineering, Leonard Hill, London (1971).

18. Data for Figure 6.2 was compiled from the following sources:
Anonymous, U.S. Nitrogen 1967-68 -- A Year of Change," Nitrogen, No. 57, pp. 19-25 (1969).

Cope, Willard C., "Ammonia Part I: Production Facilities," Chemical Industries, 64, pp. 920-925 (June, 1949).

Ammonia and Synthesis Gas, Noyes Development Corporation, Pearl River, New York, 1964.

Chemical Economics Handbook, Stanford Research Institute, Menlo Park, California, 1968.

Chemical Engineering, Plant Index, various years.

Fertilizer Trends, Tennessee Valley Authority, Muscle Shoals, Alabama, various issues.

Stumpe, J.J., The Nitrogen Industry in the United States, Tennessee Valley Authority, Office of Agricultural and Chemical Development, Division of Chemical Development, Muscle Shoals, Alabama, 1973.

Taylor, George V., "Nitrogen Processes and Facilities," in Jacob, K.D. (ed.), Fertilizer Technology and Resources in the United States, Academic Press, Inc., New York, 1953.

19. Winton, John M., "The Big Ammonia Buildup," Chemical Week, 93:18, pp. 117-126 (November 2, 1963).

20. Anonymous, "Too Much Too Soon?" Chemical Week, 76:6, pp. 75-76 (February 5, 1965).

21. Anonymous, "Petrochemicals," Chemical Week, 81:13, p. 52 (September 28, 1975).

22. Anonymous, "Market Newsletter," Chemical Week, 84:7, p. 72 (February 14, 1959).

23. Anonymous, "More Atmospheric Ammonia Storage Coming," Chemical and Engineering News, 38:34, pp. 46-47 (August 22, 1960).

24. Anonymous, "Swing to Cold NH$_3$ Saves Storage $$," Chemical Engineering, 68:4, pp. 88-90 (February 20, 1961).

25. Stanford Research Institute, Chemical Economics Handbook, Stanford Research Institute, Menlo Park, California, 1968.

26. Anonymous, "Market Newsletter," Chemical Week, 87:18, p. 104 (October 20, 1960).

27. K.M. Guthrie, "Capital and Operating Costs for 54 Chemical Processes," Chemical Engineering, pp. 140-56 (June 15, 1970).

28. Based on various issues of Hydrocarbon Processing, annual processing handbook.

29. Johnston, J., Econometric Methods, 2nd Edition, McGraw-Hill, New York, 1972, pages 249-54.

REFERENCES TO APPENDIX

1. "Midwest Ammonia Pipelines Now in Operation," Chemical and Engineering News, pp. 17-18 (January 4, 1971).

2. Burt, R.B., "Conversion from Coke to Natural Gas as Raw Material in Ammonia Production," Industrial and Engineering Chemistry, 46:12, pp. 2479-86 (December, 1954).

3. Cope, W.C., "Ammonia: Part II: Cost of Production and End-Use Pattern," Chemical Industries, 65, pp. 53-56 (July, 1949).

4. Hein, L.B., "Synthesis of Ammonia at 350 Atmospheres," Chemical Engineering Progress, 48:8, pp. 412-18 (August, 1952).

5. Tour, R.S., "The German and American Synthetic-Ammonia Plants - V," Chemical and Metallurgical Engineering, 26:8, pp. 463-465 (March 8, 1922).

6. Duff, B.S., "Economics of Ammonia Manufacture From Several Raw Materials," Chemical Engineering Progress, 51:1, pp. 12 J-16 J (January, 1955).

7. Blouin, G.M., "Effects of Increased Energy Costs on Fertilizer Production Costs and Technology," Tennessee Valley Authority, Muscle Shoals, Alabama, 1975.

8. Strelzoff, S., "Make Ammonia From Coal," Hydrocarbon Processing, 53:10, pp. 133-135 (October, 1974).

9. Buividas, L.J., J.A. Finneran, and O.J. Quartulli, "Alternate Ammonia Feedstocks," Chemical Engineering Progress, 70:10, pp. 21-35 (October, 1974).

10. Van den Berg, G.J., P. Rijnaard, and D.J. Bryne, "How Partial Oxidation Pressure Affects Ammonia Production Costs," Hydrocarbon Processing 45:5, pp. 193-197 (May, 1966)

11. Mayland, B.O., E.A. Comley, and J.C. Reynolds, "Ammonia Synthesis Gas," Canadian Chemical Processing, pp. 22-29 (August, 1954).

12. Eickmeyer, A.G., and W.H. Marshall, "Ammonia Synthesis Gas Generation By Pressure Reforming of Natural Gas," Chemical Engineering Progress, 51:9, pp. 418-421 (September, 1955).

13. de Piccioto, A., and G.C. Sweeny, "Ammonia Manufacture from Petroleum Feed-Stocks," prepared for the United Nations Centre for Industrial Development, pp. 368-377 (November, 1964).

14. Institut Francais du Petrole, "The Petrochemical Industries - Production of Nitrogenous Fertilizers," prepared for the United Nations Centre for Industrial Development, November, 1964.

15. Bresler, S.A., and G.R. James, "Questions and Answers on Today's Ammonia Plants," Chemical Engineering, 72:13, pp. 109-118 (June 21, 1965).

16. Anonymous, "Building Big for Indian Ammonia," Chemical Week, 97:15, pp. 79-80 (October 9, 1965).

17. Anonymous, "Centrifugals Cut Ammonia Costs," Hydrocarbon Processing, 45:5, pp. 179-180 (May, 1966).

18. Chemical Process Handbook, Stanford Research Institute, Menlo Park, California (1968).

19. Quartulli, O.J., and W. Turner, "Economic and Technological Factors in Giant Ammonia Plant Design," Nitrogen, No. 80, pp. 28-34 (1972).

20. Sweeney, N., "Here's What Users Pay for Ammonia," Hydrocarbon Processing, 47:9, pp. 265-8 (September, 1968).

21. Finneran, J.A., N.J. Sweeney, and T.G. Hutchinson, "Startup Performance of Large Ammonia Plants," Chemical Engineering Progress, 64:8, pp. 72-77 (August, 1968).

22. Anonymous, "Overcome Start-Up Problems - Then NH$_3$ Profits are Good," European Chemical News pp. 36-37 (June 2, 1967).

23. Douglas, J.R., J.K. Parker, and E.A. Harre, "Fertilizer Production and Distribution Costs," Commercial Fertilizer, pp. 18-23 (Sept., 1968).

24. Ernst, F.A., "Ammonia Production and Conversion Costs," Transactions of the American Institute of Chemical Engineers, 17: Part 1, pp. 1-26 (December 3, 1925).

25. Quartulli, O.J., "Check List for High Pressure Reforming," Hydrocarbon Processing, 44:4, pp. 151-162 (April, 1965).

26. The main sources of the information displayed in Figures A.5 and A.6 are:

Ammonia and Synthesis Gas, Noyes Development Corp., Pearl River, New York, 1964.

Anonymous, "U.S. Nitrogen 1967/68 -- A Year of Change," *Nitrogen*, No. 57, pp. 19-25 (1969).

Breslaner, J., "World Nitrogen Industry Survives International Crisis," *Chemical and Metallurgical Engineering*, 43:5, pp. 282-285 (May, 1936).

Chemical Economics Handbook, Stanford Research Institute, Menlo Park, California, 1968.

Chemical Engineering, Plant Index, various years.

Cope, Willard C., "Ammonia: Part I: Production Facilities," *Chemical Industries*, No. 64, pp. 920-925 (June, 1949).

Fertilizer Trends, Tennessee Valley Authority, Muscle Shoals, Alabama, various issues.

Industrial and Engineering Chemistry, 17:8, p. 772 (August, 1925).

Stumpe, J.J., *The Nitrogen Industry in the United States*, Tennessee Valley Authority, Office of Agricultural and Chemical Development, Division of Chemical Development, Muscle Shoals, Alabama, 1973.

Taylor, George V., "Nitrogen Processes and Facilities," *Fertilizer Technology and Resources in the United States*, Jacob, K.D. (ed.), Academic Press, Inc., New York (1953).